Studies in Computational Intelligence

Volume 717

Series editor

Janusz Kacprzyk, Polish Academy of Sciences, Warsaw, Poland
e-mail: kacprzyk@ibspan.waw.pl

About this Series

The series "Studies in Computational Intelligence" (SCI) publishes new developments and advances in the various areas of computational intelligence—quickly and with a high quality. The intent is to cover the theory, applications, and design methods of computational intelligence, as embedded in the fields of engineering, computer science, physics and life sciences, as well as the methodologies behind them. The series contains monographs, lecture notes and edited volumes in computational intelligence spanning the areas of neural networks, connectionist systems, genetic algorithms, evolutionary computation, artificial intelligence, cellular automata, self-organizing systems, soft computing, fuzzy systems, and hybrid intelligent systems. Of particular value to both the contributors and the readership are the short publication timeframe and the worldwide distribution, which enable both wide and rapid dissemination of research output.

More information about this series at http://www.springer.com/series/7092

Stefka Fidanova

Editor

Recent Advances in Computational Optimization

Results of the Workshop on Computational Optimization WCO 2016

 Springer

Editor
Stefka Fidanova
Department of Parallel Algorithms, Institute
  of Information and Communication
  Technology
Bulgarian Academy of Sciences
Sofia
Bulgaria

ISSN 1860-949X   ISSN 1860-9503   (electronic)
Studies in Computational Intelligence
ISBN 978-3-319-86720-5   ISBN 978-3-319-59861-1   (eBook)
DOI 10.1007/978-3-319-59861-1

Printed on acid-free paper

This Springer imprint is published by Springer Nature
The registered company is Springer International Publishing AG
The registered company address is: Gewerbestrasse 11, 6330 Cham, Switzerland

Preface

Many real-world problems arising in engineering, economics, medicine, and other domains can be formulated as optimization tasks. Every day we solve optimization problems. Optimization occurs in the minimizing time and cost or the maximization of the profit, quality, and efficiency. Such problems are frequently characterized by non-convex, non-differentiable, discontinuous, noisy or dynamic objective functions and constraints which ask for adequate computational methods.

This volume is a result of very vivid and fruitful discussions held during the Workshop on Computational Optimization. The participants have agreed that the relevance of the conference topic and quality of the contributions have clearly suggested that a more comprehensive collection of extended contributions devoted to the area would be very welcome and would certainly contribute to a wider exposure and proliferation of the field and ideas.

This volume includes important real problems such as job scheduling, wildfire modeling, parameter settings for controlling different processes, capital budgeting, data mining, finding the location of sensors in a given network, identifying the conformation of molecules, algorithm correctness, decision support system, and computer memory management. Some of them can be solved applying traditional numerical methods, but others need huge amount of computational resources. Therefore, for them is more appropriate to develop an algorithm based on some metaheuristic methods such as evolutionary computation, ant colony optimization, particle swarm optimization, and constrain programming.

Sofia, Bulgaria
April 2017

Stefka Fidanova
Co-Chair
WCO'2016

Organization

Workshop on Computational Optimization (WCO 2016) is organized in the framework of FEDERATED CONFERENCE ON COMPUTER SCIENCE AND INFORMATION SYSTEMS FedCSIS—2016

Conference Co-chairs

Stefka Fidanova	IICT (Bulgarian Academy of Sciences, Bulgaria)
Antonio Mucherino	IRISA (Rennes, France)
Daniela Zaharie	West University of Timisoara (Romania)

Program Committee

Bartl, David	University of Ostrava, Czech Republic
Bonates, Tibérius	Universidade Federal do Ceará, Brazil
Breaban, Mihaela	University of Iasi, Romania
Chira, Camelia	Technical University of Cluj-Napoca, Romania
Gonçalves, Douglas	Universidade Federal de Santa Catarina, Brazil
Hosobe, Hiroshi	National Institute of Informatics, Japan
Iiduka, Hideaki	Kyushu Institute of Technology, Japan
Krislock, Nathan	Northern Illinois University, United States
Lavor, Carlile	IMECC-UNICAMP, Campinas, Brazil
Marinov, Pencho	Bulgarian Academy of Science, Bulgaria
Muscalagiu, Ionel	Politehnica University Timisoara, Romania
Ninin, Jordan	ENSTA-Bretagne, France
Parsopoulos, Konstantinos	University of Patras, Greece
Pintea, Camelia	Tehnical University Cluj-Napoca, Romania

Roeva, Olympia	Institute of Biophysics and Biomedical Engineering, Bulgaria
Siarry, Patrick	Universite Paris XII Val de Marne, France
Stefanov, Stefan	Neofit Rilski University, Bulgaria
Stuetzle, Tomas	Universite Libre de Bruxelles, Belgium
Tamir, Tami	The Interdisciplinary Center (IDC), Israel
Zilinskas, Antanas	Vilnius University, Lithuania

Contents

Sequential Predictive Scheduling in Partitioned Data Domains

Jörg Bremer, Christian Hinrichs, Sönke Martens
and Michael Sonnenschein

Abstract Following the long-term goal of substituting conventional, fossil power generation completely with cleaner, renewable energy will consequently lead to an integration of a large share of small energy generation units imposing large problem sizes for coordination. Hardly predictable, stochastic feed-in makes the problem even harder. Predictive scheduling is a frequent task in energy grid control and has been widely studied for some decades. But, the expected huge number of entities leads to a need for new techniques reducing the computational effort for coordination. For a group of energy resources, a schedule has to be found for each single entity in the group that fulfills several objectives at the same time and resembles jointly a wanted target schedule. Considering day-ahead scenarios with 96-dimensional schedules imposes additional challenges to this already hard combinatorial problem. We explore the effects of reducing complexity by partitioning the data domain of the optimization problem for a sequential approach that integrates energy models for constraint handling directly into the optimization process. We explore the effects of different partitioning schemes and evaluate the trade-off between accuracy and effort with several simulation studies.

1 Introduction

In European countries, especially in Germany where currently a financial security of guaranteed feed-in prices is granted, the share of distributed energy resources (DER) is rapidly rising. Following the goal defined by the European Commission

J. Bremer (✉) · C. Hinrichs · S. Martens · M. Sonnenschein
University of Oldenburg, Uhlhornsweg 84, 26129 Oldenburg, Germany
e-mail: joerg.bremer@uni-oldenburg.de

C. Hinrichs
e-mail: christian.hinrichs@uni-oldenburg.de

S. Martens
e-mail: soenke.martens@uni-oldenburg.de

M. Sonnenschein
e-mail: sonnenschein@uni-oldenburg.de

© Springer International Publishing AG 2018
S. Fidanova (ed.), *Recent Advances in Computational Optimization*,
Studies in Computational Intelligence 717, DOI 10.1007/978-3-319-59861-1_1

1

[17], concepts for integration into electricity markets will quickly become indispensable for both: active power provision as well as ancillary services to reduce subsidy dependence [1, 28]. Consequently, combining smart measurement technologies for decentralized information gathering on current operational grid state, new remote control techniques, communication standards and decentralized self-organized control schemes will lead to a so called smart grid with power conditioning and control of the production and distribution of electricity managed without central control; as in the vision of [38] or similar in Europe [14].

Despite being environmentally friendly and sustainable, the increasing amount of renewable electricity generation has a major drawback. In contrast to conventional power plants, the generation from e.g. solar and wind power can neither be predicted with high accuracy nor scheduled precisely. Furthermore, as storage of electrical energy is a rather difficult and expensive task, balancing supply and demand in the grid in real-time is one of the most important functions of power system control centers. Thus, to incorporate renewables accordingly, methods have to be established that can compensate for the missing flexibility of those energy sources. For instance, controlling flexible loads to use electrical power in times of high availability (i.e. high wind or solar radiation) can help using renewable power more efficiently [31].

From an algorithmic perspective, the task of scheduling energy units can be seen as combinatorial optimization problem: For each unit (i.e. controllable load or generation), an optimal schedule has to be found such that for every time interval of a predefined planning horizon, a specific amount of electrical power (positive or negative) is assigned. A combination of schedules is optimal if the aggregated power equals a target profile that is given as defined by the use case. For example, given the inverse of a predicted feed-in time series for wind and photovoltaic power plants as target profile, an optimal schedule assignment for the controllable energy units would lead to a perfect balancing of supply and demand in the considered system at each interval of the prediction horizon. Another use case is the operation of a virtual power plant (VPP) [15, 29, 30]. A VPP is a group of individually operated energy resources, distributed over the grid and – from an control point of view – drawn together by means of communication. In this way, the group is jointly controlled and acts, seen from the outside, as a single, larger power plant.

Given a target power profile that is to be offered in an energy market, the members of the VPP must collaborate in such a way that the VPP as a whole will produce the target profile. From the outside perspective, no difference between a VPP and a classical power plant would be evident [31].

However, the schedule optimization task becomes hard to solve in the presence of device-specific restrictions. Many flexible generators and loads are controllable in principle, but at the same time have to obey specific individual constraints. For instance, a co-generation plant (i.e. a combined heat and power plant, CHP) produces thermal and electrical power simultaneously. As the generation of those two forms of power are strictly coupled within the unit and the use of the heat is subject to further restrictions such as the size of an attached thermal buffer storage, the electrical generation is severely confined as well [2, 6]. Due to such constraints, many established optimization algorithms cannot be applied to this task. For instance, meta-heuristics like evolutionary algorithms or simulated annealing are not able to cope with constraints per se and would have to be tailored specifically for the actual use case and the involved energy units.

In [4], a method has been introduced that is able to transform a problem with restrictions into a restriction-free representation using a machine learning approach. This so-called *support vector decoder model* allows generic optimization algorithms to operate in a restriction-free representation of the constrained search space of the original optimization problem. The method has been successfully applied to the schedule optimization problem [6] as well as to other problem classes from smart grid control [5, 9]. In the context of predictive scheduling, the influence of the length of the planning horizon on solution quality became apparent: Usually, the method is applied to representations of the planning horizon as a whole by interpreting feasible schedules of energy units as elements to the combinatorial problem. However, the longer the planning horizon (and the schedules, consequently), the lower the solution quality of the employed optimization algorithms. At first glance, this may seem like an inherent restriction of the problem to solve. But interestingly, preliminary experiments indicated a potentially increasing solution quality when the optimization algorithm is applied in a successive manner to sequential partitions of the planning horizon.

In [20], the potential benefit of partitioning the search space of the given combinatorial problem in the data domain in combination with sequential optimization of the individual data partitions have been explored for the first time. In this paper, the results from [20] are further extended for an more in-depth discussion of the underlying mechanisms.

In Sect. 2, the motivating optimization problem as well as the support vector decoder model are briefly recapped from previous works. Following, Sect. 3 first revisits relevant related work in the field of high-dimensionality optimization strategies before describing the introduced concept of data partitions for the considered combinatorial problem in more detail. Section 4 then evaluates the approach by employing a simulation study in the aforementioned application domain. Finally, Sect. 5 concludes the paper.

2 Methodical Background

We start with some preliminary definitions. First, let $\mathcal{U}$ be the set of DER units in the VPP and Z_U be the set of operational states of unit U. We regard the schedule of an energy unit as a vector $\boldsymbol{p} = (p_1, \ldots, p_d) \in \mathbb{R}^d$ of mean power p_i generated (or consumed) during the ith time interval. The starting time and the width of a time interval (today usually 15 min) are defined separately and have no effect on this representation. For the used support vector decoder it is advantageous to use schedules with scaled power values [10]. Scaling is done according to respective minimum (p_{min}) and maximum (p_{max}) nominal active power output (or input):

$$\rho : \mathbb{R}^d \to \mathscr{X} \subset [0, 1]^d$$

$$\boldsymbol{p} \mapsto \boldsymbol{x} = \rho(\boldsymbol{p}), \text{ with } x_i = \frac{p_i - p_{min}}{p_{max} - p_{min}}; \tag{1}$$

For this paper we go with the example of predictive scheduling for active power planning in day-ahead scenarios (not necessarily 24 h but for some given future time period).

One of the crucial challenges in operating a VPP arises from the complexity of the scheduling task due to the large amount of (small) energy units in the distribution grid [25]. In the following, we consider predictive scheduling, where the goal is to select exactly one schedule $\boldsymbol{x}_i$ for each energy unit U_i from a search space of feasible schedules with respect to a future planning horizon, such that a global objective function (e.g. a target power profile for the VPP) is optimized by the sum of individual contributions [34]. A basic formulation of the scheduling problem is given by

$$\delta \left(\sum_{i=1}^{m} \boldsymbol{x}, \boldsymbol{\zeta} \right) \to min \tag{2}$$

such that

$$\boldsymbol{x}_i \in \mathscr{F}^{(U_i)} \; \forall U_i \in \mathscr{U}. \tag{3}$$

In Eq. (2) δ denotes an (in general) arbitrary distance measure for evaluating the difference between the aggregated schedule of the group and the desired target schedule $\boldsymbol{\zeta}$. W.l.o.g., in this contribution we use the Euclidean distance $\| \cdot \|_2$. To each energy unit U_i exactly one schedule $\boldsymbol{x}_i$ has to be assigned. The desired target schedule is given by $\boldsymbol{\zeta}$. $\mathscr{F}^{(U_i)}$ denotes the individual set of feasible schedules that are operable for unit U_i without violating any (technical) constraint. Solving this problem without unit independent constraint handling leads to specific implementations that are not suitable for handling changes in VPP composition or unit setup without having changes in the implementation of the scheduling algorithm [29].

In [3] a so called support vector decoder has been introduced. Basically, a decoder is a constraint handling technique that gives an algorithm hints on where to look for feasible solutions. It imposes a relationship between a decoder solution and a feasible solution and gives instructions on how to construct a feasible solution [13]. For example, [22] proposed a homomorphous mapping between an n-dimensional hyper cube and the feasible region in order to transform the problem into an topological equivalent one that is easier to handle. In order to be able to derive such a decoder mapping automatically from any given energy unit model, [3] developed an approach based on a support vector model [10]. We will briefly describe this method.

The basic idea is to start with a set $\mathscr{X} = \{\boldsymbol{x}_i\}_n$ of feasible example schedules derived from the simulation model of an energy unit and use this sample as a stencil for the region (the sub-space in the space of all schedules) that contains only feasible schedules. The set $\mathscr{X}$ can be easily generated after a sampling method from [7]. The schedule sample is then used as a training set for a support vector based machine

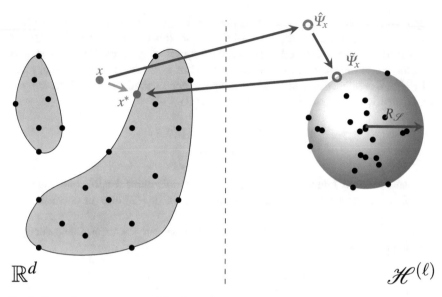

Fig. 1 General support vector model and decoder scheme for solution repair and constraint handling

learning approach [35] that derives a geometrical description of the sub-space that contains the given data (in our case: the feasible schedules). Given a set of data samples, the inherent structure of the scope of action of a unit where the data resides in can be derived as follows: After mapping the data to a high dimensional feature space by means of an appropriate kernel, the smallest enclosing ball in this feature space is determined. When mapping back this ball to data space, it forms a set of contours (not necessarily connected) enclosing the given data sample. An in-depth discussion can for example be found in [35].

At this point, the set of alternatively feasible schedules of a unit is represented as pre-image of a high-dimensional ball $\mathscr{S}$. Figure 1 shows the situation. This representation has some advantageous properties. Although the pre-image might be some arbitrary shaped non-continuous blob in $\mathbb{R}^d$, the high-dimensional representation is still a ball and thus geometrically easier to handle (right hand side of Fig. 1). The relation is as follows: If a schedule is feasible, i.e. can be operated by the unit without violating any technical constraint, it lies inside the feasible region (grey area on the left hand side in Fig. 1). Thus, the schedule is inside the pre-image (that represents the feasible region) of the ball and thus its image in the high-dimensional representation lies inside the ball. An infeasible schedule (e.g. x in Fig. 1) lies outside the feasible region and thus its image $\hat{\Psi}_x$ lies outside the ball. But we know some relations: the center of the ball, the distance of the image from the center and the radius of the ball. Hence, we can move the image of an infeasible schedule along the difference vector towards the center until it touches the ball. Finally, we calculate the pre-image of the moved image $\tilde{\Psi}_x$ and get a schedule at the boundary of the feasible region:

a repaired schedule x^* that is now feasible. We do not need a mathematical description of the original feasible region or of the constraints to do this. The decoder that does the trick is derived directly from the training set $\mathscr{X}$ generated from the respective simulation model. More sophisticated variants of transformation are e.g. given in [4]. For a detailed description of the support vector decoder approach we refer to [4]. Formally, we have a mapping (the decoder γ)

$$\gamma : [0, 1]^d \to \mathscr{F}_{[0,1]} \subseteq [0, 1]^d$$
$$x \mapsto \gamma(x) \tag{4}$$

that transforms any given (maybe in-feasible) schedule into a feasible one. Thus, we are able to transform the scheduling problem given Eq. (2) into an unconstrained formulation.

With these preliminaries in constraint handling we can now reformulate our optimization problem as

$$\delta\left(\sum_{i=1}^{m} \rho_i^{-1} \circ \gamma(x_i), \zeta \right) \to min, \tag{5}$$

where γ_i is the decoder function of unit i that produces feasible, scaled schedules from $x \in [0, 1]^d$ and ρ_i^{-1} scales them unit specific entrywise to correct active power values (inverse to Eq. (1)) resulting in schedules that are operable by that unit. Please note, that this is a constraint free formulation. With this problem formulation, many standard algorithms for optimization can be easily adapted as there are no longer any constraints (apart from a simple box constraint $x \in [0, 1]^d$) to be handled and no domain specific implementation (regarding the energy units and their operation schedules) has to be integrated. Equation (5) is used as a surrogate objective to find the solution to the constrained optimization problem Eq. (2).

Using a decoder fairly eases the implementation of a solver because no complex constraints have to be considered. On the other hand, such a decoder may introduce additional complexity into the optimization problem with this transformation. For this reason, we scrutinized the fitness landscapes of both problems (untransformed and transformed) to gain insight into the problem structure with means from standard fitness landscape analysis [36]. Indeed, our findings indicate a slightly growing in complexity by an increased ruggedness with a growing number of local minima [8]. But, this situation can be easily countered by using a heuristics that copes well with rugged non-linear problems like Simulated Annealing (SA).

Simulated Annealing [21] is an established Markov Chain Monte Carlo Method (MCMC) for non-linear optimization. It mimics a physical cooling process. In general, MCMC methods are an effective tool for statistical sampling applied to optimization problems [23]. The basic idea is a Markov Process that samples a target probability distribution $\pi(x) = \frac{1}{z} e^{-E(x)}$ with z as a problem specific normalization parameter and E measuring the error of the optimization objective. Originally, the method has been mainly applied to physical problems finding a minimum energy state and thus E is sometimes still denoted as Hamiltonian $\mathscr{H}$, e.g. in [27]. We will use the term E. In this process a new state σ_{t+1} is generated from σ_t by drawing from a proposal transition distribution $Q(\sigma_{t+1}|\sigma_t)$ [18, 26]. The new state is accepted with probability

$$A(\sigma_t \rightarrow \sigma_{t+1}) = \min\left(1, \frac{\pi(\sigma_{t+1})Q(\sigma_{t+1}|\sigma_t)}{\pi(\sigma_t)Q(\sigma_t|\sigma_{t+1})}\right). \tag{6}$$

The proposal distribution Q is a free parameter and must be adjusted to the individual problem at hand. Starting from a random initial state σ_0, the process needs a while to reach equilibrium and independence from σ_0. After this burn-in phase the samples represent the target distribution π.

In systems with deep local minima the process can be trapped without escape in reasonable time. This waiting time dilemma [39] is due to a stringent requirement for equilibrium. To escape, the process must generate subsequent states with higher energy and the probability for such a move declines roughly exponentially with the energy differences that has to be overcome. Thus, the expected waiting time for such escape grows also exponentially. For high-dimensional problems like the one that we scrutinize here, this problem is even more prevalent [39]. Several techniques have been proposed to overcome the problem of getting trapped, e.g. [12, 24, 39]; one is the concept of Simulated Annealing (SA).

SA introduces a variable temperature T into the target distribution: $\pi(x) = \frac{1}{z}e^{-E(x)/T}$. The effect is that the Markov Chain may escape local minima easier at a higher temperature. The general idea of Simulated Annealing is to interpret the fitness landscape of an optimization problem as a thermodynamic system with the objective function $E(x)$ denoting the error interpreted as the energy level of a proposed solution x. Initially, the system is at a high temperature. During the Markov process, the system is gradually cooled down to the ground state with the global energy minimum.

Algorithm 1 shows the basic flow within our SA with integrated decoder. This integration has first been proposed in [8]. By mimicking a cooling process, temporarily worse solutions are allowed – depending on temperature and difference in solution quality – in order to escape local minima. In our approach, a solution is described by two matrices X_{ij} and M_{ij} denoting for each energy unit i and for each time interval j of the schedule a scaled active power value in $[0, 1]$. In many objective scenarios, indicator values that describe the schedule with respect to different objectives might additionally be prevalent. For demonstration purposes, we stick with the single objective case here. In this sense, each row within the matrix is the schedule for one of the units. X contains schedules from the unconstrained search space (hypercube $[0, 1]^d$ not further constrained by technical issues from the units' operations). X is initialized with random values. M concurrently holds the respective feasible values generated by the support vector decoder: $M_i = \gamma_i(X_i)$. Thus, M always represents a feasible (scaled) solution to the problem.

X and M represent the genotype and phenotype of a solution respectively. In each iteration of the SA exactly one schedule x from X is randomly chosen and mutated. Modification is done at a randomly chosen element x_k by adding a random value $p \sim N(0, 1)$:

$$x_k \leftarrow \begin{cases} x_k + p - 1 & \text{if } x_k + p > 1 \\ x_k + p + 1 & \text{if } x_k + p < 0 \\ x_k + p & \text{else.} \end{cases} \qquad (7)$$

Additionally, it can be useful especially for high-dimensional schedules to allow mutations at more than one element at a time. Only this mutated schedule has to be mapped by the respective decoder in order to keep M consistent with X.

The system evolves as follows: at each temperature level T^t a Markov chain samples $E(x)$. M always represents a feasible, mutated solution that can be evaluated by Eq. (5). The new proposal solution part x^{t+1} is accepted (according to the Metropolis-Hastings criterion) with probability

$$A(x^t \rightarrow x^{t+1}) = \min\left(1, e^{\frac{-\Delta E}{T^t}}\right), \qquad (8)$$

with $\Delta E = E(x^{t+1}) - E(x^t)$. In each iteration, temperature T^t is updated with with cooling rate $\lambda \in [0, 1[: T^{t+1} \leftarrow \lambda \cdot T^t$.

Algorithm 1 Basic scheme for the Simulated Annealing step (with integrated support vector decoder).

1: $X_{ij} \leftarrow x_i \sim U(0, 1)^d, \ 1 \leq i \leq n$
2: $M_{ij} \leftarrow \gamma_i(X_i), \ 1 \leq i \leq n$
3: $\vartheta \leftarrow \vartheta_{start}$
4: **while** $\vartheta < \vartheta_{min}$ **do**
5:     choose random k; $1 \leq k \leq n$
6:     $x^* \leftarrow X_k$
7:     mutate(x^*)
8:     $M^* \leftarrow M$; $M_k^* \leftarrow \gamma_k(x^*)$
9:     **if** $e^{-\frac{E(M^*)-E(M)}{T}} > r \sim U(0, 1)$ **then**
10:         $M \leftarrow M^*$; $X_k \leftarrow x^*$
11:     **end if**
12:     $T \leftarrow$ cooling(T)
13: **end while**

A major advantage of this approach is the anytime property: at any time, a feasible solution exists. The Markov chain may evolve in $[0, 1]^{d \cdot n}$ without taking care of technical constraints of the individual energy units. The decoder guarantees (apart from minor inaccuracies that might easily be corrected [4]) the feasibility of the solution.

3 Partitioning the Search Space

By employing the support vector decoder approach in combination with a heuristic solver for the optimization problem as described in the previous section, we are able to solve the scheduling problem for energy units efficiently without needing to adapt any part of the process to unit-specific properties such as technical constraints. The whole process is visualized in Algorithm 2. The resulting matrix M comprises m rows and d columns, where the ith row vector represents the chosen schedule for energy unit U_i (for the remaining symbol definitions refer to Sect. 2).

Algorithm 2 Predictive Scheduling

1: $m \leftarrow$ amount of energy units
2: $n \leftarrow$ sample size per energy unit
3: $d \leftarrow$ length of planning horizon
4: **for all** energy unit $U_i \in \mathcal{U}$ **do**
5:     $s_i \leftarrow$ predicted state of U_i at the beginning of the planning horizon
6:     **repeat**
7:         initialize simulation model for U_i with s_i
8:         simulate feasible schedule of length d
9:     **until** $\mathcal{F}^{(U_i)}$ contains n feasible schedules
10:     scale sample $\mathcal{F}^{(U_i)}$ using ρ_i
11:     calculate support vector model $\mathcal{S}_i$
12:     build support vector decoder γ_i
13: **end for**
14: **return** $M \leftarrow \left(\text{solve } \delta \left(\sum_{i=1}^{m} \rho_i^{-1} \circ \gamma(x_i), \zeta \right) \to min \right)$

In the considered application domain, predictive planning is commonly done for *day-ahead* planning horizons, i.e. d corresponds to 24 h with a schedule resolution of 15 min. In our problem formulation, this yields a 96-dimensional search space for each energy unit. Due to the curse of dimensionality [16], this may introduce significant negative effects. For instance, with larger problem dimensions, the required amount of training data for the support vector model increases exponentially [37]. This affects both the generation of feasible schedule samples via simulation, as well as learning the support vector models from these samples. Moreover, solving the optimization problem itself gets more time-consuming due to combinatorial explosion. Finally, as the support vector decoder model is based on approximation, mapping accuracy deteriorates with larger dimensions. This may lead to infeasible schedules being misleadingly recognized as feasible.

According to [32], strategies to circumvent the curse of dimensionality in such a case can be categorized as follows:

- *Decomposition:* Given that the problem is separable, decomposition subdivides the problem into smaller parts that are easier to solve.
- *Screening:* Less significant and redundant decision variables/dimensions are pruned from the problem description in order to reduce dimensionality.

- *Mapping:* The problem is mapped to a representation comprising less dimensions. For example, by exploiting correlations between variables in the original space, a mapping can be designed that yields a correlation-free space with less dimensions.
- *Space Reduction:* Using expert knowledge, parts of the search space are excluded from optimization.
- *Visualization:* An expert prunes insignificant parts of the search space using visualization techniques for high-dimensional data. In contrast to *Space Reduction*, this is done interactively during the optimization process.

For the considered support vector decoder approach, the strategies Screening, Visualization, and Space Reduction with expert's help are inappropriate, as they rely on specific knowledge about the individual problem instance to solve, which contradicts the main motivation for our approach. Because neighbouring values in the unit schedules are often quite similar (i.e. the gradient between two time intervals is usually rather small) and thus show some correlation, Mapping might be applicable. After optimization, however, the resulting low-dimensional power profile would have to be inversely mapped to a feasible high-dimensional schedule again, which would introduce further problems.

Finally, Decomposition offers a viable solution. We cannot split the problem along the m axis with respect to the result matrix M in Algorithm 2 (i.e. by optimizing over disjunct sets of energy units), because in each time step along the d axis, the schedule selections of *all* participating units have to be regarded in order to minimize δ. On the other hand, the problem formulation might allow us to optimize over each time step along the d axis independently: If the employed distance measure δ is a *metric*, it gets minimal if the individual distances along the d axis are minimal. This holds true for the Euclidean distance $\| \cdot \|_2$ we are using in this paper. Therefore, from the optimization point of view, the given problem seems to be separable along the d axis. Formally, we define such a *partitioning* of the search space as

$$\pi : \mathbb{N}^2 \to \mathbb{N}$$
$$(d, j, k) \mapsto l = \pi(d, j, k), \tag{9}$$

where $l = \pi(d, j, k)$ denotes the length of the jth partition along the d axis. The parameter k may hold arbitrary implementation-specific values (cf. the equidistant partitioning below). For a partitioning to be valid, the concatenation of all generated partitions must yield the whole planning horizon:

$$\sum_{j=1}^{\infty} \pi(d, j, k) = d \tag{10}$$

Moreover, for convenience we require

$$\forall i : \pi(d, j, k) = 0 \Rightarrow \pi(d, j + 1, k) = 0, \tag{11}$$

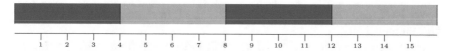

Fig. 2 Equidistant partitioning for $d = 16$ and $k = 4$

i.e. as soon as the partitioning function yields the first zero partition, every following partition must be zero as well. Using this rather general definition of π, we may now define different partitioning strategies. For example, the *equidistant partitioning* subdivides the planning horizon into k partitions of equal size

$$\pi_{\text{eq}}(d, j, k) = \begin{cases} \lceil \frac{d}{k} \rceil & \text{if } j \leq k \ \wedge \ j \leq d \bmod k, \\ \lfloor \frac{d}{k} \rfloor & \text{if } j \leq k \ \wedge \ j > d \bmod k, \\ 0 & \text{else.} \end{cases} \tag{12}$$

Figure 2 shows an example for this partitioning with $d = 16$ and $k = 4$. There are many other possible partitioning strategies, ranging from simple arithmetic fragmentations to more sophisticated strategies involving expert knowledge about the use case at hand (i.e. the structure of the target profile or the δ function). A particular promising approach is the *entropy partitioning*, which exploits the entropy in the feasible schedule samples to determine intervals of high versus low flexibility in the units' scopes of actions, and partitions the search space accordingly.

Algorithm 3 shows the possible approach for achieving such an entropy-based partition. The entropy is determined for each time interval within the horizon and compared with the previously calculated mean entropy; scaled by a parametrizable factor. The length of a partition then reflects the deviation of the local from the mean entropy. Figure 3 shows an example result. Regions with a high variety in possible

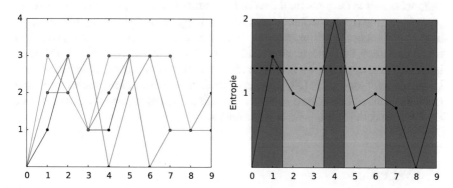

Fig. 3 Entropy partitioning for $m = 0.8933$ and $k = 1.5$ and thus threshold $m \cdot k = 1.34$; number of time interval on x-axis. The *left* figure shows exemplarily one training schedule for each device (real power on y-axis); the *left* figure shows the resulting entropy per time interval and the partitioning according to the given threshold

power levels and thus with a high flexibility result in shorter partitions due to the high entropy there.

Nevertheless, in order to remain maximally independent from such expert knowledge, we go with the example of equidistant partitioning in the remainder of this paper.

Algorithm 3 Partitioning by entropy

---

function $p(d, i)$

1: $m \leftarrow \frac{\sum_{t=1}^{d} - \sum_j P_{t,j} \cdot \log_2(P_{t,j})}{d}$ mean entropy over all time intervals t

2: $s \leftarrow \sum_{j=1}^{i-1} p(d, j)$

3: $l \leftarrow 0$;

4: $sum \leftarrow 0$;

5: **while** $sum < m \cdot k \wedge s + l < d$ **do**

6:     $l++$

7:     $sum \leftarrow sum - \sum_j P_{s+l,j} \log_2(P_{s+l,j})$ sum up entropy until threshold reached

8: **end while**

9: **return** l length of partition

In order to implement a partitioning scheme like e.g. the equidistant partitioning for the Simulated Annealing approach to high dimensional predictive scheduling, we have to extend Algorithm 2. Special care has to be taken regarding the simulation of feasible schedules: Originally, in Algorithm 2, each simulation model was initialized with the state of the energy unit right at the beginning of the planning horizon, and was executed for d time steps, such that each schedule sample exactly covers the planning horizon. Using partitions, however, schedule samples cannot be generated beforehand for the whole planning horizon. In order to identify a unit's flexibility for a certain partition, the exact state of the unit at the beginning of this partition has to be known. Thus, before being able to process a partition, we have to assign fixed schedules to the units for the preceding partition. As a consequence, the overall process ranging from schedule simulation to solving the optimization problem has to be executed for each partition separately. This ensures that, after the process finished for all partitions, the concatenated result schedules are feasible overall. On the other hand, with this approach we achieve a reduction of the design space (without expert knowledge as proposed in [32]) as every subsequent optimization process is already tackled to a fixed operational state of each unit at the beginning of a partition. The resulting process is visualized in Algorithm 4.

4 Evaluation

The objective of this paper is to explore the potential benefit of partitioning the search space of the given combinatorial problem in the data domain, followed by sequential optimization of the individual partitions. In the previous section, a partitioning

---

Algorithm 4 Predictive Scheduling with Partitioning

1: $m \leftarrow$ amount of energy units
2: $n \leftarrow$ sample size per energy unit
3: $d \leftarrow$ length of planning horizon
4: **for all** energy unit $U_i \in \mathscr{U}$ **do**
5:    $s_i \leftarrow$ predicted state of U_i at the beginning of the planning horizon
6: **end for**
7: $j \leftarrow 1$
8: $k \leftarrow$ implementation specific value
9: **while** $\pi(d, j, k) \neq 0$ **do**
10:    **for all** energy unit $U_i \in \mathscr{U}$ **do**
11:       **repeat**
12:          initialize simulation model for U_i with s_i
13:          simulate feasible schedule of length $\pi(d, j, k)$
14:       **until** $\mathscr{F}^{(U_i)}$ contains n feasible schedules
15:       scale sample $\mathscr{F}^{(U_i)}$ using ρ_i
16:       calculate support vector model $\mathscr{S}_i$
17:       build support vector decoder γ_i
18:    **end for**
19:    $M^j \leftarrow \left(\text{solve } \delta \left(\sum_{i=1}^{m} \rho_i^{-1} \circ \gamma(x_i), \zeta \right) \rightarrow min \right)$
20:    **for all** energy unit $U_i \in \mathscr{U}$ **do**
21:       run simulation model for U_i using schedule x_i
22:       $s_i \leftarrow$ predicted state of U_i after running x_i
23:    **end for**
24:    $j \leftarrow j + 1$
25: **end while**
26: **return** $M \leftarrow [M^j]$

---

framework has been introduced for this purpose, along with a detailed description of the according optimization process chain. In order to evaluate the proposed approach with respect to the objective, a simulation study has been conducted.

4.1 Simulation Setup

Following the considered example use case, we set up a simulated virtual power plant for active power planning in day-ahead scenarios, comprising co-generation units with an $800\,\ell$ thermal buffer store each. We used the simulation model of an EcoPower CHP as described in [6]. For each of those devices, the thermal demand for a four-family house during winter was simulated. The devices were operated in heat driven operation and thus primarily had to compensate the simulated thermal demand. Additionally, after shutting down, a device would have to stay off for at least two hours. However, due to their thermal buffer store and the ability to modulate the electrical power output within the range of $[1.3,\ 4.7]\,\text{kW}$, the devices still have some flexibility available.

For the generation of feasible schedule samples, a *successive sampling strategy* was employed: Instead of guessing whole schedules and checking feasibility afterwards (using a device's simulation model), which leads to large rejection rates, a period-wise guessing in combination with partial feasibility checks is applied repeatedly to construct feasible schedules in a successive manner, cf. [7]. Preliminary experiments indicated 200 as an adequate size for $\mathscr{F}^{(U_i)}$, so we set $n = 200$ in the present study.

The planning horizon was set to $d = 96$ time intervals, i.e. 24 h in 15 min resolution, which is a common use case in the application domain. As motivated in the previous section, we employ the equidistant partitioning function π_{eq} in this study. Regarding the parameter k, which defines the length of the partitions and thus inversely determines the number of partitions to be generated according to (12), several experiments with $k \in [1, 96]$ have been conducted. For instance, $k = 1$ yields 96 partitions of length 1, while $k = 96$ corresponds to a single partition of length 96, i.e. no partitioning at all. This way, the influence of a partitioning on the optimization can be explored in a structured manner.

While k represents the primary influence factor in our study, other parameters may cause relevant interaction effects. Here, especially the magnitude of the problem size along the m axis (i.e. the number of energy units in the VPP, cf. Sect. 3) is of particular interest, as it affects the problem complexity for each partition likewise. Similarly, different target profiles ζ have to be examined with respect to the units' available flexibilities. For example, a target profile might turn out to be easily realizable due to well matching schedule options in the units' search spaces, or vice-versa. The question arises whether this influences the potential benefit of a partitioning, and how a partitioning should be done in order to gain optimal results.

In all experiments, we used the Simulated Annealing solver as outlined in Algorithm 4. Each examined parameter configuration was simulated 100 times, so that the results can be interpreted with statistical soundness.

4.2 Results

The evaluation focuses on solution quality, which is calculated as remaining error after optimization:

$$\delta \left(\sum_{i=1}^{m} \rho_i^{-1}(x_i), \zeta \right), \quad x_i \in M \tag{13}$$

where M denotes the $m \times d$ schedule matrix after all partitions have been processed (line 26 in Algorithm 4). In the following, results are visualized as box-charts, where the box spans from the upper to the lower quartile of the data. The median is shown as horizontal line within a box, whereas the whiskers span over $1.5 \times$ the interquartile range. Outliers are illustrated by circle markers.

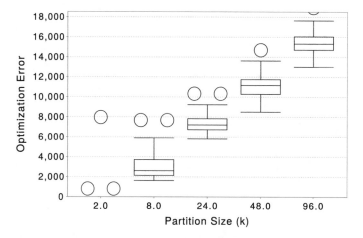

Fig. 4 Remaining optimization error for different partition sizes

First of all, the general influence of different k values (i.e. different partition sizes) is examined. As already stated in the introduction, preliminary experiments indicated a potentially increasing solution quality when the optimization algorithm is applied in a successive manner to sequential partitions of the planning horizon. For a more thorough analysis, we conducted 100 simulations for each $k \in \{2, 8, 24, 48, 96\}$. The results are visualized in Fig. 4. The optimization error clearly decreases with more and thus smaller partitions (from right to left in the figure). Comparing the extreme points, the partitioning even allows approaching the theoretical optimum $\delta = 0$ when the partitions are generated as small as possible ($k = 2$: despite a few outliers, the box is squashed to a single line at $\delta = 0$), while the no-partitioning case yields the worst results most of the time ($k = 96$).

These results support our hypothesis strikingly, but they originate from a single experiment configuration only: On the one hand, a fixed number of energy units was involved, $m = 10$. On the other hand, the target profile ζ was generated by aggregating randomly chosen sample schedules (one for each energy unit) at the beginning of each experiment run. This way, ζ formed an "easy" target, because the energy units were able to approach it optimally in principle. In the following, we will vary this configuration in these two aspects, in order to gain more insights into the involved effects.

4.2.1   Interaction with the Number of Energy Units

In the considered application use-case of predictive scheduling for active power planning in day-ahead scenarios, virtual power plants may comprise different amounts of energy units, depending on e.g. regional conditions. From the optimization point

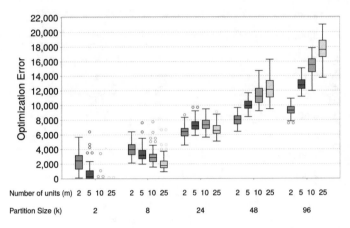

Fig. 5 Remaining optimization error for different partition sizes and varying amounts of energy units

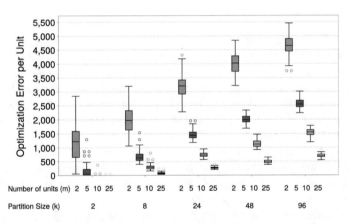

Fig. 6 Remaining optimization error per energy unit for different partition sizes and varying amounts of energy units

of view, this corresponds to the problem size along the m axis. In a partitioned setting (i.e. $k < d$), each subproblem is of size $m \times k$. Hence, m affects the problem complexity for each partition likewise. To reveal possible interactions with the magnitude of k, the previous experiment with $k \in \{2, 8, 24, 48, 96\}$ was repeated for $m \in \{2, 5, 10, 25\}$. Figure 5 visualizes the results. Similar to Fig. 4, the optimization error generally decreases with smaller partitions. Within each block, however, different effects with respect to the number of units m are visible: For the case of small partitions, the optimization error is lower with larger values of m, while this trend reverses for large partitions. As this is based on the absolute error, which is naturally different for varying magnitudes of m, Fig. 6 complementarity shows the same results against the normed optimization error with respect to the number of units, i.e. the remaining error per energy unit. Here, the trend towards a lower error for small partitions is again clearly visible, whereas the magnitude of m results in a change of the slope for this trend. Concluding, m seems to affect the problem complexity only as a whole, and does not seem to interact with the partition size k.

4.2.2 Dependencies on the Target Profile

In active power planning, usually an application-specific target profile is given. For instance, in day-ahead energy market scenarios, a target profile would be chosen such that the economic outcome of the VPP is maximized. In contrast, in supply-demand-matching scenarios, the target profile might be e.g. a constant zero value, such that the considered set of energy units (flexible producers and consumers) can be treated as autonomous energy-wise. While it is advisable to configure VPP and target profile in a matching way, so that the latter is actually a feasible target for the former, not all target profiles are equally easy to realize.

In our study, we abstract from application-specific scenarios as follows. As a first step, a feasible target can be formed by aggregating randomly chosen sample schedules (one for each energy unit). This way, the existence of the theoretical optimum ($\delta = 0$) is guaranteed. We denote this type of target with ζ_0. To generate more difficult target profiles in an easy but structured way, ζ_0 can simply be shifted in magnitude:

$$\zeta_i = \zeta_0 + i \tag{14}$$

Please note that ζ is a vector, and the summation is performed element-wise. Matching the size of the considered VPP in the present study, we choose values for i between 0 kW and ± 1 kW in the following experiment, in order to deviate the target profile from "easy to solve optimally" towards "hard to solve optimally". Thus, as in the previous section, the original experiment with $k \in \{2, 8, 24, 48, 96\}$ was repeated for all ζ_i with $i \in \{-1, -0.5, -0.25, 0, 0.25, 0.5, 1\}$ in kW. The results are presented in Fig. 7. Similar to the results from Fig. 5, the general trend of better optimization results with smaller partitions is visible. The case $k = 2$, $i = -1$ is an exception. Here, the optimization was not able to find a feasible schedule at all in the available time. This is due to the very low values in the target profile in combination with a large number of partitions: Due to the independent optimization of individual partitions, the simulated CHP units stay off at the beginning of the planning horizon until the thermal buffer stores are exhausted. At that point in time, however, thermal demand exceeds the available power from the CHPs, so that no feasible schedule can be found anymore (Fig. 8).

Figure 9 shows the situation for both of these extreme partitions: for partition size 2 (depicted on the left), corresponding to the maximum number of subsequent optimization problem and thus corresponding to the shortest visible (manageable)

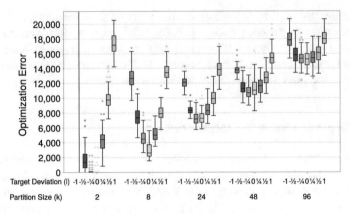

Fig. 7 Remaining optimization error for different partition sizes and varying target profile deviations

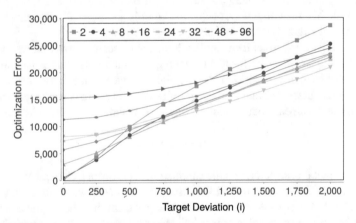

Fig. 8 Remaining optimization error for varying target profile deviations (horizontal axis) and different partition sizes (individual data series)

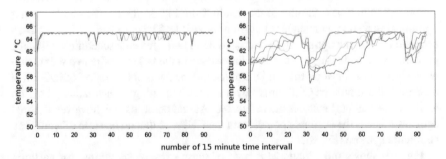

Fig. 9 Example for premature exhaustion of flexibility along the example of remaining thermal buffer capacity. On the *left* No remaining flexibility is preserved due to a partition size of 2. On the *right* Buffer flexibility is properly exploited due to an anticipatory optimization with partition size 96

time frames. For partition size 96 (on the right), the optimization has full information (just one single optimization problem). As can be observed in the latter case, temperature flexibility is not exhausted if the optimization has no timely notice about a need for preserving remaining flexibility in succeeding time intervals. Of course, this leads to an unavoidable degradation in solution quality but results in at least feasible solutions.

With larger partitions, the effect is not present, as the optimization can act anticipatory towards feasibility (i.e. by choosing schedules that lead to a poor optimization error, but in turn form a feasible solution). This effect indicates that a strong partitioning can yield better optimization results if enough flexibility is present, but might also lead to infeasible solutions in extreme cases.

In addition to the general trend regarding the value of k, a u-shaped course can be seen within each configuration of the same partition size. In other words, solution quality seems to deteriorate with larger deviations from ζ_0, which is not surprising at all. In order to focus on the interaction between these two effects, Fig. 8 visualizes the results in a transposed way, i.e. the deviation i is visualized along the horizontal axis, while the partition sizes k are presented as line charts.

For visualization purposes, the shown data comprises mean values only. Furthermore, in this experiment a larger amount of configurations was examined: $k \in \{2, 4, 8, 16, 24, 32, 48, 96\}$ and $|i| \in \{0, 0.25, \ldots, 2\}$. The results reveal an interesting relationship: For smaller target deviations, configurations with smaller partitions yield superior optimization results. In contrast, for larger target deviations, larger partition sizes yield better results. In summary, the last experiment supports our previous hypothesis: With enough flexibility in a given problem configuration (in terms of feasible solution combinations with respect to the fitness function), the solver significantly benefits from a partitioning. On the other hand, in more difficult problem formulations (i.e. with less flexibility in terms of feasible solutions), the solver cannot cope with a large number of independent partitions.

Finally, the sensitivity of algorithm parametrization has been scrutinized. Figure 10 shows a result for the example of the mutation rate. For the used Simulated Annealing, basically two parameters have to be chosen. Other parameters apply to the search space and decoder model and are discussed for example in [11, 19]. For the SA, a mutation rate and a cooling rate have to be fixed. For the cooling rate that determines how fast the process converges (with the risk of premature convergence with too fast cooling), empirically a value of 0.9999 has been found which can be applied to a wide range of problems and partition sizes. The mutation rate denotes the number of mutated schedule elements that are modified in each iteration. The impact of the mutation rate is described in Fig. 10. Here, the following relation can be observed: for small partitions sizes smaller mutation rates are advantageous for large partitions sizes higher rates pay off.

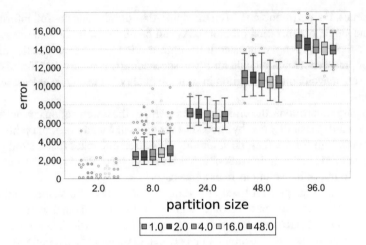

Fig. 10 Sensitivity of the mutation rate (varying from 1 to 48) of the result quality for different partition sizes

5   Conclusion

The objective of this paper was to explore the potential benefit of partitioning the search space of a combinatorial problem in the data domain using the example of predictive scheduling as a smart grid use case. We combined the partitioning approach with a sequential optimization solving each partition successively. As predictive scheduling is a constrained optimization problem, simulation models [33] of different energy units have been integrated directly in the process for handling individual search spaces and operational constraints.

Several methods to cope with the challenge of high dimensionality in optimization problems have been proposed in the past. A good overview on methods for computationally expensive black-box functions (as might be the case when using simulation models for computing objectives) is e.g. given in [32]. Our approach is a mixture of design space reduction and decomposition into sub-problems. To achieve this we have to introduce simulation models as black-boxes into the optimization process for sequencing. Introducing this sequence of independently solvable sub-problems reduces the overall computationally effort and at the same time reduces the design space so that modeling is more accurate and optimization effort is reduced [32]. At the same time this reduction leads to a limited choice especially for later sub-problems. Sub-space parts of the design space may be missed [32]. On the other hand, with our method, we may focus on the whole sub-space at once without a need for subsequent refinement like in other methods [32].

Our results support the hypothesis of an increasing solution quality when applying the optimization algorithm in a successive manner to sequential partitions of the planning horizon. For our experiments we mainly used a simulated annealing approach as solver although our results can be generalized to other solvers. In general, any solver benefits for partitioned data domains in predictive scheduling if a problem configuration contains enough flexibility in terms of feasible solution combinations. With decreasing flexibility, additional complexity induced by a growing number of partitions prevails.

So far all simulations have been done with scenarios regarding predictive scheduling. Additional use cases like load balancing can be easily adapted by exchanging the objective functions, as the problem structure is similar to predictive scheduling. Future work will concentrate on methods to classify the situation at hand in order to automatically decide on appropriate partition of the combinatorial problem.

References

1. Abarrategui, O., Marti, J., Gonzalez, A.: Constructing the active European power grid. In: Proceedings of WCPEE09, Cairo (2009)
2. Arteconi, A., Hewitt, N., Polonarac, F.: Domestic demand-side management (DSM): role of heat pumps and thermal energy storage (TES) systems. Appl. Thermal Eng. 51(1–2), pp. 155–165 (2013). doi:10.1016/j.applthermaleng.2012.09.023
3. Bremer, J., Sonnenschein, M.: A distributed greedy algorithm for constraint-based scheduling of energy resources. In: Ganzha, M., Maciaszek, L.A., Paprzycki, M. (eds.) FedCSIS, pp. 1285–1292 (2012)
4. Bremer, J., Sonnenschein, M.: Constraint-handling for optimization with support vector surrogate models – a novel decoder approach. In: Filipe, J., Fred, A. (eds.) ICAART 2013 – Proceedings of the 5th International Conference on Agents and Artificial Intelligence, SciTePress, Barcelona, Spain, vol. 2, pp. 91–100 (2013). doi:10.5220/0004241100910100
5. Bremer, J., Sonnenschein, M.: Estimating shapley values for fair profit distribution in power planning smart grid coalitions. In: Klusch, M., Thimm, M., Paprzycki, M. (eds.) Multiagent System Technologies - 11th German Conference, MATES 2013, Koblenz, Germany, September 16–20, 2013, Proceedings. Lecture Notes in Computer Science, vol. 8076, pp. 208–221. Springer, Berlin (2013). doi:10.1007/978-3-642-40776-5_19
6. Bremer, J., Sonnenschein, M.: Model-based integration of constrained search spaces into distributed planning of active power provision. Comput. Sci. Inf. Syst. 10(4), 1823–1854 (2013). doi:10.2298/CSIS130304073B
7. Bremer, J., Sonnenschein, M.: Sampling the search space of energy resources for self-organized, agent-based planning of active power provision. EnviroInfo, Berichte aus der Umweltinformatik, pp. 214–222. Shaker, Germany (2013)
8. Bremer, J., Sonnenschein, M.: Parallel tempering for constrained many criteria optimization in dynamic virtual power plants. In: 2014 IEEE Symposium on Computational Intelligence Applications in Smart Grid, CIASG 2014, Orlando, FL, USA, December 9–12, 2014, pp. 51–58. IEEE (2014). doi:10.1109/CIASG.2014.7011551
9. Bremer, J., Lehnhoff, S.: Decentralized coalition formation in agent-based smart grid applications. Highlights of Practical Applications of Scalable Multi-Agent Systems. The PAAMS Collection. Communications in Computer and Information Science, vol. 616, pp. 343–355. Springer, Berlin (2016)
10. Bremer, J., Rapp, B., Sonnenschein, M.: Encoding distributed search spaces for virtual power plants. In: IEEE Symposium Series on Computational Intelligence 2011 (SSCI 2011), Paris, France (2011). doi:10.1109/CIASG.2011.5953329
11. Bremer, J., Rapp, B., Sonnenschein, M.: Encoding distributed search spaces for virtual power plants. In: Computational Intelligence Applications in Smart Grid (CIASG), 2011 IEEE Symposium Series on Computational Intelligence (SSCI), Paris, France (2011). doi:10.1109/CIASG. 2011.5953329

12. Brown, S., Head-Gordon, T.: Cool walking: a new markov chain monte carlo sampling method. J. Comput. Chem. **24**(1), 68–76 (2003). doi:10.1002/jcc.10181
13. Coello Coello, C.A.: Theoretical and numerical constraint-handling techniques used with evolutionary algorithms: a survey of the state of the art. Comput. Methods Appl. Mech. Eng. **191**(11–12), 1245–1287 (2002). doi:10.1016/S0045-7825(01)00323-1
14. Colak, I., Fulli, G., Sagiroglu, S., Yesilbudak, M., Covrig, C.F.: Smart grid projects in Europe: current status, maturity and future scenarios. Appl. Energy **152**, 58–70 (2015). http://dx.doi.org/10.1016/j.apenergy.2015.04.098
15. Coll-Mayor, D., Picos, R., Garciá-Moreno, E.: State of the art of the virtual utility: the smart distributed generation network. Int. J. Energy Res. **28**(1), 65–80 (2004). doi:10.1002/er.951
16. Donoho, D.L.: High-dimensional data analysis: the curses and blessings of dimensionality. In: Aide-memoire of a Lecture at AMS Conference on Math Challenges of the 21st Century (2000)
17. European Parliament & Council: Directive 2009/28/ec of 23 april 2009 on the promotion of the use of energy from renewable sources and amending and subsequently repealing directives 2001/77/ec and 2003/30/ec
18. Hastings, W.K.: Monte carlo sampling methods using markov chains and their applications. Biometrika **57**(1), 97–109 (1970). doi:10.1093/biomet/57.1.97
19. Hinrichs, C., Bremer, J., Sonnenschein, M.: distributed hybrid constraint handling in large scale virtual power plants. In: IEEE PES Conference on Innovative Smart Grid Technologies Europe (ISGT Europe 2013). IEEE Power & Energy Society (2013). http://www-ui.informatik.uni-oldenburg.de/download/Publikationen/HBS13.pdf
20. Hinrichs, C., Bremer, J., Martens, S., Sonnenschein, M.: partitioning the data domain of combinatorial problems for sequential optimization. In: Ganzha, M., Maciaszek, L., Paprzycki, M. (eds.) 9th International Workshop on Computational Optimization, Proceedings of the 2016 Federated Conference on Computer Science and Information Systems, Gdansk (2016, in press)
21. Kirkpatrick, S., Gelatt, C.D., Vecchi, M.P.: Optimization by simulated annealing. Science **220**(4598), 671–680 (1983). doi:10.1126/science.220.4598.671
22. Koziel, S., Michalewicz, Z.: Evolutionary algorithms, homomorphous mappings, and constrained parameter optimization. Evol. Comput. **7**, 19–44 (1999). doi:10.1162/evco.1999.7.1.19
23. Li, Y., Protopopescu, V.A., Arnold, N., Zhang, X., Gorin, A.: Hybrid parallel tempering and simulated annealing method. Appl. Math. Comput. **212**(1), 216–228 (2009). doi:10.1016/j.amc.2009.02.023
24. Marinari, E., Parisi, G.: Simulated tempering: a new Monte Carlo scheme. Europhys. Lett. **19**(6) (1992)
25. McArthur, S., Davidson, E., Catterson, V., Dimeas, A., Hatziargyriou, N., Ponci, F., Funabashi, T.: Multi-agent systems for power engineering applications–Part I: concepts, approaches, and technical challenges. IEEE Trans. Power Syst. **22**(4), 1743–1752 (2007). doi:10.1109/TPWRS.2007.908471
26. Metropolis, N., Rosenbluth, A.W., Rosenbluth, M.N., Teller, A.H., Teller, E.: Equation of state calculations by fast computing machines. J. Chem. Phys. **21**(6), 1087–1092 (1953). doi:10.1063/1.1699114
27. Müller, A., Schneider, J.J., Schömer, E.: Packing a multidisperse system of hard disks in a circular environment. Phys. Rev. E **79**, 021102 (2009). doi:10.1103/PhysRevE.79.021102
28. Nieße, A., Lehnhoff, S., Tröschel, M., Uslar, M., Wissing, C., Appelrath, H.J., Sonnenschein, M.: Market-based self-organized provision of active power and ancillary services: an agent-based approach for smart distribution grids. In: Complexity in Engineering (COMPENG), 2012, pp. 1–5 (2012). doi:10.1109/CompEng.2012.6242953
29. Nieße, A., Beer, S., Bremer, J., Hinrichs, C., Lünsdorf, O., Sonnenschein, M.: Conjoint dynamic aggregation and scheduling for dynamic virtual power plants. In: Ganzha, M., Maciaszek, L.A., Paprzycki, M. (eds.) Federated Conference on Computer Science and Information Systems - FedCSIS 2014, Warsaw, Poland (2014). doi:10.15439/2014F76
30. Nikonowicz, Ł.B., Milewski, J.: Virtual power plants – general review: structure, application and optimization. J. Power Technol. **92**(3) (2012). http://papers.itc.pw.edu.pl/index.php/JPT/article/view/284/492

31. Palensky, P., Dietrich, D.: Demand side management: demand response, intelligent energy systems, and smart loads. IEEE Trans. Ind. Inform. **7**(3), 381–388 (2011). doi:10.1109/TII. 2011.2158841

32. Shan, S., Wang, G.G.: Survey of modeling and optimization strategies to solve high-dimensional design problems with computationally-expensive black-box functions. Struct. Multidiscip. Optim. **41**(2), 219–241 (2010). doi:10.1007/s00158-009-0420-2

33. Sonnenschein, M., Appelrath, H.J., Canders, W.R., Henke, M., Uslar, M., Beer, S., Bremer, J., Lünsdorf, O., Nieße, A., Psola, J.H., et al.: Decentralized provision of active power. In: Smart Nord - Final Report. Hartmann GmbH, Hannover (2015)

34. Sonnenschein, M., Hinrichs, C., Nieße, A., Vogel, U.: Supporting renewable power supply through distributed coordination of energy resources. In: Hilty, L.M., Aebischer, B. (eds.) ICT Innovations for Sustainability. Advances in Intelligent Systems and Computing, vol. 310, pp. 387–404. Springer, Berlin (2015). doi:10.1007/978-3-319-09228-7_23

35. Tax, D.M.J., Duin, R.P.W.: Support vector data description. Mach. Learn. **54**(1), 45–66 (2004). doi:10.1023/B:MACH.0000008084.60811.49

36. Vassilev, V.K., Fogarty, T.C., Miller, J.F.: Information characteristics and the structure of landscapes. Evol. Comput. **8**(1), 31–60 (2000). doi:10.1162/106365600568095

37. Verleysen, M., François, D.: The curse of dimensionality in data mining and time series prediction. Computational Intelligence and Bioinspired Systems. Lecture Notes in Computer Science, vol. 3512, pp. 758–770. Springer, Berlin (2005). doi:10.1007/11494669_93

38. Vinay Kumar, K., Balakrishna, R.: Smart grid: advanced metering infrastructure (AMI) & distribution management systems (DMS). Int. J. Comput. Sci. Eng. **3**(11) (2015)

39. Wong, W.H., Liang, F.: Dynamic weighting in Monte Carlo and optimization. Appl. Math. Proc. Nat. Acad. Sci. **94**, 14220–14224 (1997)

Wildfire Optimizations in Modeling and Calibrations for Bulgarian Test Cases

Nina Dobrinkova

Abstract In this article we are going to present the optimizations that has been done through different types of models applied on wildland fires for Bulgarian test cases. We will present approaches where meteorological data along with terrain specific relief and vegetation coverage are modeled in a way to present credible scenarios for wildland propagation used for calibration purposes of the different cases. This work aims to prove that the used modeling tools can be used as decision support mechanism for the responsible authorities when it is combined with field observations and simulated propagation scenarios. In conclusion we will give as working example a web-based system in USA which with adaptations can be applicable for Bulgarian conditions.

1 Introduction

The work presented in this paper is a year's long efforts which have been started because of an accident that happened in Pirin Mountain near by the city of Razlok. In the year 2003 a helicopter with water tank flew very low to a running fire trying to suppress it dumping water quantities over the mountainous area. Unfortunately the engine oxygen has been vacuumed because of the flames, which caused helicopter's crash with four people crew that died that day [1]. This accident was very problematic for the Bulgarian society. That is why in the Bulgarian scientific community has been launched in the beginning of 2007 a pilot Ph.D. program dedicated to the wildland propagation and its modeling opportunities as first attempts for computer based simulations on wildland fires propagation in Bulgaria.

In 2007 small team from Bulgarian Academy of Sciences (BAS) started adaptation of a US model, which was running in parallel mode. The model was called WRF-Fire (in 2010 renamed SFIRE). The input data for the model had to be first collected for a specific test area and second preprocessed for model calibration.

N. Dobrinkova (✉)
Institute of Information and Communication Technologies – Bulgarian
Academy of Sciences, acad. Georgi Bonchev bl. 2, Sofia, Bulgaria
e-mail: nido@math.bas.bg

© Springer International Publishing AG 2018

S. Fidanova (ed.), *Recent Advances in Computational Optimization*,
Studies in Computational Intelligence 717, DOI 10.1007/978-3-319-59861-1_2

The area of interest for the BAS team at first was nearby Sofia, where idealized case has been run with the model functionalities in order to set the model parameters. No real fire has been on this area. So no calibration has been done. However the second attempt was run for a real test case wildland fire, which has occured near by the village of Leshnikovo, region of Harmanli in the past. For the second test case we tried as much as possible to do a model set up repeating the steps of fire propagation as it was described by the forester's department in Harmanli.

In this paper we will show the basis of the mathematical calculations and optimizations outlined from the research efforts and the achieved results as final outcomes. This work is ongoing and more optimizations will occur in future.

2   WRF-Fire (SFIRE) Mathematical Basis

The mathematical background of the WRF-Fire model (SFIRE) is as position in the (x, y) plane. The model is semi-empirical and it represents the spread of the fire in direction of the fire line. This is the so called Rothermel modified formula. The burning region is represented as Ω for time t, which is represented with the point coordinates (x, y). The formula itself is:

$$\tilde{S} = \min\{B_0, R_0 + \phi_w + \phi_s\}, \tag{1}$$

where B_0 is the fire spread against the wind direction, R_0 is the fire spread in absence of wind, $\phi w = a(\vec{v} \cdot \vec{n})b$ is the wind correction and $\phi s = d\nabla z \cdot \vec{n}$ is the terrain correction, $\vec{v}$ is wind, ∇z is terrain variable along the normal $\vec{n}$ of the fire line, a, b и d are constants. In this case WRF-Fire use:

$$S = \begin{cases} 0, \ if \ \tilde{S} < 0 \\ S_{\max}, \ if \ \tilde{S} > S_{\max} \\ \tilde{S}, \ if \ 0 \le \tilde{S} \le S_{\max} \end{cases}, \tag{2}$$

where $S_{\max}$ is max fire spread. After the burning materials are burnt the model decrease them in the points (x, y) exponentially and that is represented with the formula:

$$F(x, y, t) = F_0(x, y)e^{-(t-t(x,y))/W(x,y)}_i, \tag{3}$$

where t is the time, t_i is the time for the burning, F_0 is the initial quantity of the burning materials (before they started to burn) and W(x,y) does not depend on the time, but from the burning materials. The heat transfer released by the fire, is represented in the atmosphere model as layer above the surface, which is situated in height [2]. The burning material quantity is represented by:

$$\Phi = -A(x, y)\frac{\partial}{\partial t}F(x, y, t) \, . \tag{4}$$

This representation is needed because the atmosphere model WRF, does not support border values for heat transfer. The coefficients B_0, R_0, S_{max}, a, b, d, W and A, which describe the burning materials are measured in laboratory with experiments. For every surface point in the plane the coefficients of the burning materials are represented using the 13 Anderson categories [3]. These categories are developed for US originally and they have been defined by usage of the different sea levels on the surface. WRF-Fire (SFIRE) has internally representation of every category and all additional characteristics, which gives opportunity for modifications when the fire is outside US.

WRF-Fire use also level-set functions for the spread of the fire [4]. This approach set as function $\psi = \psi(x, y, t)$, which define for Ω subregions using the rule:

$$\Omega(t) = \{(x, y) \in \Omega : \psi(x, y, t) < 0\}. \tag{5}$$

These subregions are burned and the fire line is defined as curve:

$$\Gamma(t) = \{(x, y) \in \Omega : \psi(x, y, t) = 0\}. \tag{6}$$

The function $\psi(x, y, t)$ satisfy the equation:

$$\frac{\partial \psi}{\partial t} + S(x, y)|\nabla \psi| = 0, \tag{7}$$

which can be solved numerically.

Formulas (1)–(7) are general description how mathematically the fire spread is represented inside the WRF-Fire (SFIRE) model. In the beginning the atmosphere model is interpolating the wind in order to get into the bigger domain of the atmosphere the fire changes. Afterwards is applied numerical method for the level-set function. The next step is to apply quadratic formulas for evaluation of the burnt material. In parallel it is evaluated also the released heats transfer into the atmosphere layers. The last step gives atmospheric change and that trigger the repetition of the model to starts again.

3 Experimental Results with WRF-Fire (SFIRE)

The experimental results which were obtained after evaluation of the WRF-Fire (SFIRE) model can be presented in brief summary in this section.

The first experimental run of the model was for an ideal case, nearby Sofia city using coordinates and information for the village of Leshnikovo. The used version of the model was WRF-Fire v.3.2 for the simulation. We did domain with size 4 by

4 km, with horizontal resolution of 50 m, for the atmosphere mesh, the used grid was 80 by 80 cells and with 41 vertical levels from ground surface up to 100 hPa. We didn't use nesting to keep the ideal case as basic as possible in order to evaluate the model capacity and set up the needed initial conditions for a Bulgarian runs.

The domain, was set with a location 4 km west from village Zheleznitsa in the south-east part of Sofia city region. The domain was covering the lower part of the forests of Vitosha mountain.

The ignition line was set in the center of the domain and the ignition line was 345 m long. The model at that stage of its development did not allow us to use ignition from a point, because the atmospheric model did not cover such narrow measurements. The ignition in parallel has been set to start 2 s after the simulation has begun. The results from this first simulation gave us idea how the model can be initialized and what the input data will be if we start calibration simulations with real wildland fire test case.

That is why we selected from the national data base in the ministry of forests, food and agriculture, a fire which has been burning in the period 14–17 August 2009 located near by the city of Harmanli. For the initialization of the model with the real test case we had to use algorithm for implementation of the real data in a way that WRF-Fire (SFIRE) was able to recognize it. We set for these purposes two domains: the first was covering area of 48 km^2 with resolution 300 m (160×160). This domain was producing boundary and initial meteorological conditions for the inner domain and in this domain were no fire simulations.

The inner domain was located in the middle of the coarse domain. The resolution in Domain 2 was set as 60 m and the area covered is 9.6 km^2 (161×161). Domain 2 was centered on the fire ignition line and it was covering the areas of villages Ivanovo, Leshnikovo and Cherna Mogila. This area was located in South-East Bulgaria close to the Bulgarian-Greece border nearby Harmanli city.

The first data source which was very important was the meteorological input. We used US NCEP Global Analyses data for meteorological background input: The data was with 1×1 degree grid resolution covering the entire globe, the time resolution was 6 h. With this data we could simulate all over the world but with resolution of around 100 km.

The next data set, which was needed for input in WPS (WRF pre-processor) was the topography data. The standard topo-data used in WPS was USGS 30 s resolution global data set (GTOPO30), but because terrain elevation is very important for correct fire behavior we used much more detailed data for the area of Harmanli (this data was available also for the whole country of Bulgaria) from USGS / SRTM 3 s data (http://eros.usgs.gov; Shuttle Radar Topography Mission (SRTM) Finished Grade Data) in order to be used in WPS it had to be converted in a special format. The data received from the server was through a GIS raster format (DTED format *.dt1) in Lat/Long format, datum WGS84. The open source Quantum GIS (www.qgis.org) interpolate the missing data of the raster (if any) with simple linear interpolation and then we could change the projection to the one we were going to use in WRF – Lambert

Conformal Conic (ref_lat = 41.84, ref_lon = 25.936, truelat1 = 41.82, truelat2 = 41.86, stand_lon = 25.936). After the reprojecting, we exported the raster in the new format – GeoTiff. This format was used in the WPS program "convert_geotiff". The procedure we followed was explained in details at [5]. The resulting fails after convert_geotiff were used by WPS/geogrid. In order to have fire behavior, we needed fuel data for the simulation domain. Because there was no such data, for our test case area we had to choose information about the vegetation from the project Corine Landcover (funded by the European Environment Agency and the member states). This landcover data for Bulgaria was with 100 m resolution and 25 ha minimum mapping unit. (http://www.eea.europa.eu/data-and-maps/data/corine-land-cover-2006-raster). The downloaded data (it gives options for GIS vector or raster formats) was used to create fuel data, based on the land cover information. The applied classification was the 13th Anderson classes from 1982 [6]. Also orthophoto data from the geoportal of The Ministry of Regional Development and Public Works (MRDPW) of Bulgaria was used to make even more detailed the available fuel data for the domains of interest, namely Harmanli region. All rivers, lakes, villages and forest areas have been vectorized using the orthophoto images combined with CORINE2006. The resulting fail was a GIS vector shape fail with very high accuracy of representation of non burning areas like rivers and lakes, and areas with high burning fuel level, like woods.

Following the description in [5] we got the intermediate fails which WPS could use for topography and fuels. Along with the provided global datasets with WPS we placed the newly created fails in the WPS working directory. The rest of the procedures made input fails for WRF part of the model described in the WRF on-line tutorial (http://www.mmm.ucar.edu/wrf/OnLineTutorial/index.htm). The only difference was in the geogrid program, where the output fail had 2 extra varaibles – NFUEL_CAT and ZSF. NFUEL_CAT for the 13th fuel categories done based on [6] with the detailed topography. The result burnt simulated area compared to the real one can be seen on the Figs. 1 and 2.

The simulation result on Fig. 1 has been done on a supercomputer at the University of Denver by distant connection. In Table 1 the simulation outcomes were presented according to the number of the cores used.

With this simulations for the test site nearby Harmanli city has been elaborated a methodology for collection, processing and implementation of real data for test sites on Bulgarian territory. The selected model was having as input meteorological data, DEM and only 13 FBFMs (Fire Behavior Fuels Models) which led to the idea that we can experiment also with other different models like BEHAVE Plus and FARSITE for our next steps.

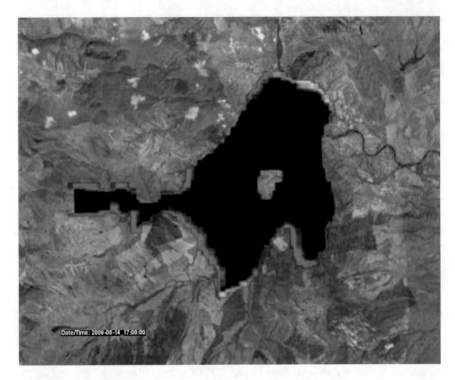

Fig. 1 The simulated burnt area

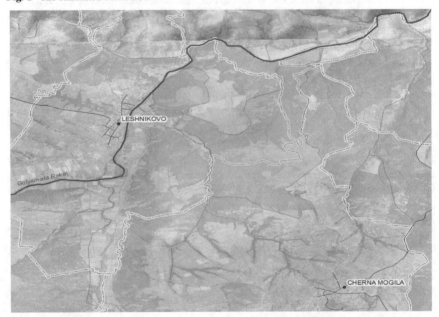

Fig. 2 The real fire burnt area

Table 1 The time required for the simulation presented in seconds depending on the number of processors running the parallel execution of processes showing that in 120 cores the simulations run as fast as real time. Everything above this cores is performing faster than real fire propagation

Cores	6	12	24	36	60	120	240	360	480	720	960	1200
Fire line propagation in km.	1.91	1.08	0.50	0.34	0.22	0.13	0.08	0.06	0.06	0.04	0.10	0.04
Region 1	6.76	7.05	2.90	2.06	1.20	0.73	0.45	0.32	0.26	0.23	0.24	0.17
Region 2	0.00	0.00	0.00	0.02	0.02	0.04	0.04	0.06	0.06	0.08	0.07	0.15
Total sec. which is the coeff. for real time	10.59	9.21	3.91	2.75	1.64	0.99	0.61	0.44	0.37	0.31	0.44	0.26

4 Experimental Implementation of BEHAVE Plus and FARSITE Simulations in the Test Cases of Zlatograd, Madan and Nedelino Municipal Areas in Bulgaria

In the framework of bilateral cooperation program between Greece and Bulgaria 2007–2013 the BAS team had the opportunity to work in the Zlatograd forestry department located on the territories of Zlatograd, Madan and Nedelino municipal areas in south – central part of Bulgaria. The study area was the territorial state-owned forestry department with its headquarters in Zlatograd. This department covers an area of 33,532 ha, where 31,856 ha are state forests. Most forests are in early to mid-serial succession stages, with only small amounts of mature to old forest. Stand age was highly variable, ranging from 20 to 80 yrs; most stands range between 35 to 50 yrs with the average being 46 yrs. Average stem stock is $140 \, m^3 \, ha^{-1}$. The average forest canopy cover was 81%.

In terms of climate, the region is part of the continental-Mediterranean climatic region, south-Bulgarian climatic sub-region and East Rodopi mountain low climate region. The average annual temperature for the area is 10.8 °C, with a maximum temperature in July of 20.6 °C and minimum temperature in January of −0.8 °C, indicating moderate summers and relatively mild winters. Extreme values of annual average maximum and minimum temperatures could be respectively 17.1 and 4.9 °C, the monthly maximum estimations are in the range in August (28.9 °C) and in January (−3.9 °C). Average annual rainfall reaches 1000 mm. Maximum precipitation amounts for the period from April to October range from 10.0 mm for 5 min to 46.3 mm for 60 min and 59.7 mm for more than 60 min. The average annual relative humidity is 75% which is an indication of good growing conditions; maximum relative humidity values of 85% which occur in November. The approximate relative humidity less than or equal to 30% was estimated about 13–15 days per year.

The data we were working on was about fifteen wildfires that occurred in 2011 to 2012 within the Zlatograd municipal territory and it was provided by the Zlatograd forestry department; this data included vegetation type, area burned (in decares where 10 decares = 1 hectare), date, and start and end hours of the fire event (Table 2). These wildfires burned in a variety of vegetation types and were more than likely started

Table 2 Wildfire for the period 2011–2012. Information provided by the Zlatograd Forestry Department

Fire No.	Vegetation type	Burned area in decares	Date of occurrence	Hour of start	Hour of end
1	Durmast	3.0	25 March 2012	1330	1530
2	Beechwood	5.0	29 March 2012	1400	1800
3	Scotch pine	1.0	16 June 2012	1500	1700
4	Scotch pine	7.0	6Aug. 2012	1640	1950
5	Scotch pine	5.0	6 Aug. 2012	1710	2130
6	European black pine	4.0	27 Aug. 2012	1200	1600
7	Scotch pine	3.0	5 Sept. 2012	1400	2030
8	Scotch pine	6.0	6 Sept. 2012	1400	1930
9	Scotch pine	2.0	6 Oct. 2012	1600	2320
10	Scotch pine	1.0	16 March 2011	1310	1400
11	Scotch pine	1.0	5 April 2011	1715	1900
12	Scotch pine	1.0	10 April 2011	1130	1530
13	Grassland	3.0	30 Aug. 2011	1400	1800
14	Scotch pine	4.0	12 Sept. 2011	1230	1900
15	Scotch pine	1.0	15 Sept. 2011	1600	1830

by humans to clear agricultural debris or prepare fields, based on the proximity to villages. Paper maps from the forestry department identified the ignition location and final fire shape; this data was digitized in a GIS which allowed each ignition point to be viewed with background orthophotos and the spatial Zlatograd vegetation classification showing pre-fire vegetation (Table 2).

After we collected and located the forest fires we did runs with BehavePlus point based prediction system in order to analyze fire growth and behavior for homogeneous vegetation with static weather data. We used standard fuel models developed for US and we evaluated which fuel models were best able to produce estimates of fire behavior and growth in BehavePlus similar to those observed on each of the fifteen fires. In addition to fuel model, BehavePlus requires inputs for weather, fuel moisture, slope, and duration of the burning period. We obtained weather data for each fire from TV Met, a private company in Bulgaria, which provided the ability to calculate fine dead fuel moisture values. Due to the paucity of available weather data in Bulgaria, we had to assume that weather recorded for the weather station closest to each particular fire is consistent with weather experienced on the wildfire. We estimated live herbaceous and live woody fuel moisture values based on the expected phenological stage for the time of year that the fire occurred. To estimate slope, we first acquired a 30 m resolution digital elevation model (DEM) from the National Institute of Geophysics, Geodesy, and Geography in Bulgaria, then subsequently calculated the average slope for each fire using standard geospatial processing in ArcGIS (ESRI 2010). Burn period length for each fire was obtained from the Zlatograd forestry department data (Table 2).

Based on initial BehavePlus results using standard fuel models, custom fuel models were developed for some vegetation types not well represented by the US fuel models. Custom fuel models were developed for native durmast oak and grass as well as one of the Scotch pine sites by modifying fuel loading parameters to better match local vegetation and reflect the lack of woody debris in the understory.

Following evaluation of fuel models with BehavePlus, we then performed analyses in FARSITE, a spatial fire growth system that integrates fire spread models with a suite of spatial data and tabular weather, wind and fuel moisture data to project fire growth and behavior across a landscape. We defined our test landscapes using a 500 m buffer zone around each of the fifteen Zlatograd fires.

Input for FARSITE consists of spatial topographic, vegetation, and fuels parameters compiled into a multi-layered "landscape file" format. Topographic data required to run FARSITE include elevation, slope, and aspect. Using the aforementioned 30 m DEM, we calculated an aspect layer, and then clipped elevation, aspect, and slope rasters to the extent of our fifteen test landscapes. Required vegetation data include fuel model and canopy cover. Fuel models within the 500 m buffered analysis area for each individual fire were assigned based on our BehavePlus analyses; fuel model assignments were tied to the dominant vegetation for each polygon based on the Zlatograd forestry department's vegetation data. Canopy cover values were visually estimated from orthophoto images and verified with stand data from the Zlatograd forestry department. Additional canopy variables (canopy base height, canopy bulk density, and canopy height) that may be included in the landscape file were omitted, as these variables are most important for calculating crown fire spread or the potential for a surface fire to transition to a crown fire. None of the fifteen fires analyzed experienced crown fire.

Tabular weather and wind files for FARSITE were compiled using the weather and wind data from TV Met, Bulgarian meteorological company that included hourly records for the purposes of our research interests. Tabular fuel moisture files were created using the fine dead fuel moisture values calculated for the BehavePlus analyses for 1-h timelag fuels. The 10-h fuel moisture value was estimated by adding 1% to the 1-h fuel moisture and the 100-h fuel moisture was generally calculated by adding 3% to the 1-h fuel moisture. The live fuel moisture values previously estimated for BehavePlus analyses were used to populate live herbaceous and live woody moisture values.

All simulations performed in FARSITE used metric data for inputs and outputs. An adjustment value was not used to alter rate of spread for standard fuel models, rather custom fuel models were created. Crown fire, embers from torching trees, and growth from spot fires were not enabled.

As an example of one of our successful FARSITE runs, we present the results from a single wildfire that burned in grassland vegetation, for which we developed custom fuel models. This fire occurred on August 30, 2011, starting at 1400 and ending around 1800, and burned a total area of 0.3 ha. We used the following input parameters to model this small grassland fire in FARSITE:

Fuel moisture values: 6 (1-h), 7 (10-h), 9 (100-h), 45 (live herbaceous), and 75% (live woody);

Daily maximum temperatures: 17–21 °C;

Daily minimum relative humidity: 24–50%;

Winds: generally from the west-southwest at $1 - 2\,k\,h-1$

The fire size as calculated using FARSITE was 0.5 ha, which seems reasonable considering the modeled size would not have included the suppression actions that most likely occurred given the close proximity of a village to this fire Fig. 3.

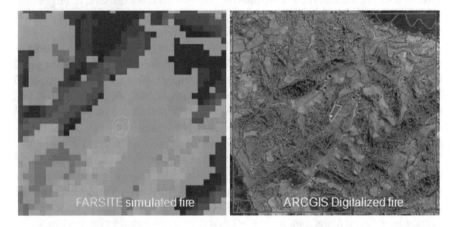

Fig. 3 FARSITE run for a grassland fire, where size of the fire is very close to the real one, but the shape is different, because of wind information discrepancies

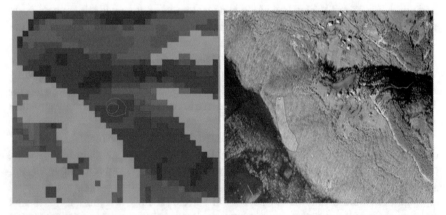

Fig. 4 FARSITE run for wildfire that burned in a beechwood forest, where size of the fire is very close to the real one, but the shape is different, because of wind information discrepancies

Table 3 Summary of the working FBFMs for the territories in Zlatograd forestry department with conclusions in which cases which type of fuel model can be applied

Vegetation type	Possible fuel models	Logic/Assumptions
Scots pine *(Pinus sylvestris)*	188 (often used for ponderosa pine) 183-modified	Ponderosa pine (pinus ponderosa) may be a suitable western US proxy. Otherwise, probably a modified 183 (TL3) to increase rate of spread and flame lengths.
Black pine/Acacia *(Pinus nigra/Acacia)*	161 183-probably modified	FBFM 161 works best whwn the understory is dominated by an herbaceous understory including forbs and grasses (it is dynamic). Creating a custom fuel model starting from FBFM 183 is another solution, to increase the rate of spread and flame lengths. Using FBFM 165 would assume ladder fuels to be present and will probably overpredict rate of spread and flame lengths.
Beechwood *(Fagus sylvatica)*	182/186 (dormant season fire) 161 (growing season fire)	FBFM 182 or 186(or a custom FBFM) may be used when a fire is mostly burning through hardwood (round leaf) litter. FBFM 186 tends to have much higher rate of spread and flame lengths than 182. FBFM 161 is dynamic and may be used during the growing season when a fire would be expected to burn through the understory vegetation.
Durmast *(Quercus dalechampii)*	182/186 (dormast season fire) 161 (growing season fire)	FBFM 182 or 186 (or a custom FBFM) may be used when a fire is mostly burning through hardwood (round leaf) litter. FBFM 186 tends to have much higher rate of spread and flame lengths than 182. FBFM 161 is dynamic and may be used during the growing season when a fire would be expected to burn through the understory vegetation.
Grasslands	101 (may be best for grazed pasture) 102 (ungrazed pasture) Custom FBFM (lower ROS and FL than FBFM 101)	Assumes no irrigation. Rate of spread and flame length drastically change depending on chosen FBFM.

An example of another fire we modeled in FARSITE using standard fuel models was a fire that occurred on March 29, 2012 in a beechwood forest. This fire burned for a total of four hours, starting at 1400 and ending around 1800, and burned a total area of 0.5 ha. Wind speeds were variable throughout the burning period as they were quite high during the early afternoon but tapered off throughout the day. In this case we used the following input parameters in FARSITE:

Fuel moisture values: 3 (1-h), 4 (10-h), 5 (100-h), 40 (live herbaceous) and 70% (live woody);

Daily maximum temperatures: 7–10 °C;

Daily minimum relative humidity: 36–40%;

Winds: generally from the north-northeast at 10−2 k h−1

The projected fire size from FARSITE was 0.9 ha. Based on the close proximity of a village to the fire location (Fig. 4) it is quite reasonable to assume that local residents responded to the fire in a volunteer capacity; these suppression actions could not be accounted for in the FARSITE analysis. Decreasing winds through the afternoon may have significantly helped suppression activities.

From this two modeled fires we were able to estimate that the standard fuel models established for US can be run also for Bulgarian cases, but only after a good calibration with real case studies where vegetation and past events are well observed. However FARSITE and BehavePlus provided reasonable outputs for future work in the field of fire behavior fuel modeling on the Bulgarian territory. As a result of the calculations with the simulations of the Fire Behavior Fuel Models we can have as summary Table 3 with the used custom and non custom FBFMs working as calibrated values for Bulgarian wildfire test cases.

The work performed with the FBFMs on the territory of Zlatograd forestry department has been published in details in the articles [7–10] where more information can be seen.

5 Potential for Implementation of the Simulated Results in Web-Based ICT Application

Natural disasters and in particular wildfires are a major problem for many European societies threatening human lives and properties, often with disastrous impacts, particularly at the wildland-urban interface. The reduction of wildfire hazard, but mainly the reduction of damages and impacts induced by wildfires, requires integrated approaches and new practices in wildland fire management, such as prescribed burning and suppression fire, together with increasing the awareness and preparedness through knowledge (education/training) and technology transfers.

In Europe the management of natural disasters is mostly within the Public Administration sector with a very small contribution from the private sector. In particular the management of forest fires is under the responsibility and within the competences of specially qualified and organized state agencies and bodies such as the Fire Departments, Forest Fires' Fighting Corps and in some cases the Forest Service itself.

To fulfill their mission, the personnel of these organizations have the necessary education and training (Formal Training). They are the professionals of forest fire fighting. In the management of natural disasters and in particular forest fires, a network of volunteers in the framework of Civil Protection Agencies is also involved, having different theoretical background and operational training levels (Informal Training). The training of these groups is often general information around the mechanisms and behavior of forest fires, or demonstrations of the use of the equipment used for the management of forest fires.

In most cases, volunteers and the professionals limit their action to logistical tasks or even spontaneous and often not well planned actions that sometimes could cause serious issues on personal safety.

That is why it becomes evident the need of creation of a system for a continuous training targeted towards the volunteer's groups and all involved professionals in wildfire suppression activities, in accordance to their needs. The most effective way to create such a system in rich forest vegetation areas and frequent episodes of forest fires, is to exchange experiences and good practices on the subject with more developed groups.

For example in Bulgaria, according to the Act of the Minister of the Interior (2009), the Civil Protection Directorate-General (DG Civil Protection) became part of the Ministry of the Interior of the Republic of Bulgaria and is responsible for performing tasks related to prevention and preparedness, management, reaction and recovery in case of natural and man-made disasters. Volunteers can participate in intervention operations together with professional firefighters, under the decision of the municipal council, and after training.

Quite the same is the situation in Greece the other south-east member state of EU affected often by wildfires as Bulgaria. According to the Act of Civil Protection (Act 3013/2003) in Greece, the General Secretariat for Civil Protection (GSCP), under the Ministry of Citizen Protection is tasked to work out prevention plans and programmes for all kinds of risks (natural and technological (CBRN included), in cooperation with all the relevant authorities at national, regional and local levels. GSCP is the authority to register Voluntary Organisations, while the National Register is listing today 240 Volunteers' Organizations.

That is why the management of natural disasters and in particular wildfires need a network of volunteers and professional firefighters to be involved, at the Local Authorities' level, having different theoretical background and operational training levels. There is a need to create a continuous and open system for the training of the volunteers and professionals, so that their participation in wildfire suppression activities to be as effective and safe as possible.

A framework for vocational training and preparedness of the volunteers and the professionals who operate within the civil protection mechanisms in general need support also from the nowadays ICT tools. Such tools already operate on the territory of USA [11], where the main idea is to have web-based decision support tools for the fire analysts who navigate the volunteer groups and the professional firefighters on the fields. In general the main idea of the so called Wildfire Decision Support System (WFDSS) is to collect information from the field by any type of observers, which is transmitted to the Fire analysts. These analysts have on their disposal information about the terrain in GIS formats (Geographical Information Systems Layers), mete-orological conditions in the area and the predefined burning materials - fuel models represented by the available 13 or 40 FBFMs [6, 12] or Custom Fuel models if no available fuels from the standard ones is not suitable.

All that information than can be calculated by a model which has in its origin mathematical, physical and chemical representation of the different fire spread rate depending on the type of the fire e.g. surface, crown, fire acceleration or spotting, which is included in the crown fires models.

In Fig. 5 is presented a general structure how the information in cases of wildfires can be collected, processed and then returned back with respective scenarios of fire spread to the people on the field. However this kind of decision support systems is not well developed outside US.

There are attempts for children education on how natural disasters can affect everyday life of people and such initiatives are described in [13, 14].

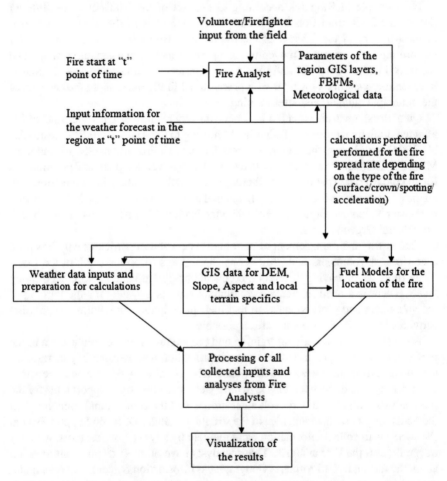

Fig. 5 General scheme of receiving information from the field by person who spot a wildfire and its process of getting supported through web-based Dession support tools such as WFDSS [11]

6 Conclusion

The presented paper was having as main aim to provide a broader view on the tested modeling options for the Bulgarian wild land fires and the achieved results. There are still a lot of issues to be solved in the data collection and processing phase. The accuracy of the meteorological inputs is still not that well developed as a network in the test areas. However the achieved results after all computations give promising options for future implementation of this modeling tools for more operational use as described the web-based decision support tool which is nowadays available in USA. Such tools and information is crucial when it comes to big wildfire events and volunteer groups work together with professionals on the field.

Acknowledgements This work has been partially supported by the Bulgarian Academy of Sciences Program for support of young researchers No: ДФНП-95-A1 and to the National Science Fund of the Bulgarian Ministry of Education, Youth and Science under Grant FNI I02/20.

References

1. http://news.ibox.bg/news/id_1888536796 (In Bulgarian)
2. Patton, E.G., Coen, J.L.: WRF-Fire: a coupled atmosphere-fire module for WRF. In: Preprints of Joint MM5/Weather Research and Forecasting Model Users Workshop, Boulder, CO, June 22–25, pp. 221–223. NCAR (2004). http://www.mmm.ucar.edu/mm5/workshop/ws04/Session9/PattonEdward.pdf
3. Anderson, H.E.: Aids to determining fuel models for estimating fire behavior. USDA Forest Service, Intermountain Forest and Range Experiment Station, Research Report INT-122 (1982). http://www.fs.fed.us/rm/pubsint/intgtr122.html
4. http://ccm.ucdenver.edu/wiki/Jan_Mandel/Blog/2010_Dec_2011_Jan
5. http://www.openwfm.org/wiki/How_to_run_WRF-Fire_with_real_data#Downloading_high_resolution_elevation_data
6. Anderson, H.E.: Aids to determining fuel models for estimating fire behavior, USDA, For. Serv. Gen. Rep. INT-122, Intermt. For and Range Exp. Stn. Ogden, Utah, pp. 1–22 (1982)
7. Dobrinkov, G., Dobrinkova, N.: Input data preparation for fire behavior fuel modeling of Bulgarian test cases (Main Focus on Zlatograd Test Case). In: 10th International Conference on "Large-Scale Scientific Computations" LSSC'15, Sozopol, 8–12 June 2015. Lecture Notes in Computer Science, vol. 9374, pp. 335–342. Springer, Germany (2015). doi:10.1007/978-3-319-26520-9 ISSN 0302-9743, ISSN 1611-3349 (electronic), ISBN: 978-3-319-26519-3
8. Dobrinkova, N., Hollingsworth, L., Heinsch, F.A., Dillon, G., Dobrinkov, G.: Bulgarian fuel models developed for implementation in FARSITE simulations for test cases in Zlatograd area. In: Wade, D.D., Fox, R.L. (eds.), Robinson, M.L. (Comp) (E-proceeding: http://www.treesearch.fs.fed.us/pubs/46778) Proceedings of 4th Fire Behavior and Fuels Conference, 18–22 February 2013, Raleigh, NC and 1–4 July 2013, St. Petersburg, Russia, pp. 513–521. International Association of Wildland Fire, Missoula, MT (2014)
9. Dobrinkov G., Dobrinkova N.: Wildfire behavior modeling data preparation for FARSITE simulations in Bulgarian test cases. In: 5th International Conference on Cartography & GIS & Seminar with EU Cooperation on Early Warning and Disaster/Crisis Management, 15–21 June 2014, Riviera, Bulgaria, Proceedings, vol. 2, pp. 763–770 (2014). ISSN:1314-0604
10. Dobrinkova, N., Dobrinkov, G.: Farsite and WRF-Fire models, Pros and Cons For Bulgarian Cases. In: 9th International Conference on "Large-Scale Scientific Computations" LSSC'13,

Sozopol 3–7 June 2013. Lecture Notes in Computer Science, vol. 8353, pp. 382–389. Springer, Germany (2014). ISBN: 978-3-662-43879-4

11. http://wfdss.usgs.gov/wfdss/WFDSS_Home.shtml
12. Scott, J.H., Burgan, R.E.: Standard fire behavior fuel models: a comprehensive set for use with Rothermel's surface fire spread model. USDA Forest Service, Rocky Mountain Research Station General Technical Report RMRS-GTR-153, Fort Collins, CO (2005)
13. Bandrova T., Kouteva, M., Pashova, L., Savova, D., Marinova, S.: Conceptual framework for educational disaster centre "Save the Children Life". In: The International Archives of the Photogrammetry, Remote Sensing and Spatial Information Sciences, ISPRS Geospatial Week 2015, 28 Sep–03 Oct 2015, La Grande Motte, France, vol. XL-3/W3, pp. 225–234 (2015)
14. Bandrova, T., Zlatanova, S., Konecny, M.: Three-dimensional maps for disaster management. ISPRS Ann. Photogramm. Remote Sens. Spatial Inf. Sci. **I–4**, 245–250 (2012). doi:10.5194/isprsannals-I-4-245-2012 (Melbourne, Australia)

Process Control with the Variability Constraints

Paweł Drąg and Krystyn Styczeń

Abstract In the article a new approach to process control with descriptor constraints has been presented. The main result is a method, which enables us to obtain consistent initial conditions. It has been proposed, that variability constraints can be used to define the consistent initial values for the index-1 differential-algebraic (DAEs) process. The variability constraints have an important practical application and have never been considered previously.

Keywords Optimal control · DAE systems · Variability constraints

1 Introduction

The mathematical models of large-scale real-life technological processes can be derived using appropriate physical laws describing observed physical phenomena [3, 4]. As a result a general system of differential-algebraic equations (DAEs) can be obtained. It means, that a state of the process can be described in a more general way, than only by purely dynamical equations (ODEs). This generalized approach can be observed in modern modeling methodologies, where the system should not be limited to the specified equations type [5]. In the other words, the mathematical model should be consisted of known relations. This approach indicates, that any additional manipulations on system equations cannot be done [6, 7].

Moreover, vision-monitoring systems, as well as image processing algorithms can be used supervise the process [1, 2]. The general scheme of the process supervised by a camera was presented on the Fig. 1. An important task in process control with the camera-based monitoring system can be connected with is a high changing rate of observed parameters. If some process trajectories are changing too fast, then applied

P. Drąg (✉) · K. Styczeń
Department of Control Systems and Mechatronics, Wrocław University of Science
and Technology, Wybrzeże Wyspiańskiego 27, 50-370 Wrocław, Poland
e-mail: pawel.drag@pwr.edu.pl

K. Styczeń
e-mail: krystyn.styczen@pwr.edu.pl

© Springer International Publishing AG 2018
S. Fidanova (ed.), *Recent Advances in Computational Optimization*,
Studies in Computational Intelligence 717, DOI 10.1007/978-3-319-59861-1_3

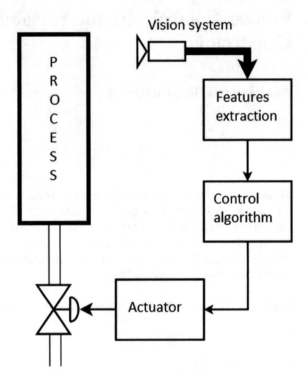

Fig. 1 The process supervised by a camera-based monitoring system

monitoring systems may become useless. Therefore, this work is concentrated on *variability constraints*, which can be treated as a new type of the process constraints. In the context of the process control, the variability constraints have been discussed together with the other differential-algebraic model equations.

The article is constructed as follows. In the Sect. 2 a control problem of DAE system without variability constraints was considered. The consistent initial conditions was discussed in Sect. 3. Then, in Sect. 4 the variability constraints were introduced, as well as a new method for control of DAE processes with the variability constraints was designed. Finally, a possible application of the presented considerations in the camera-based monitoring systems was proposed in Sect. 5.

2 Control with Differential-Algebraic Constraints

Let us consider the optimal control problem with the differential-algebraic constraints. In the considered task, the DAE system takes a form of a descriptor model

$$F_d(\mathbf{y}_d(t), \mathbf{u}(t), \mathbf{p}, t) = 0, \tag{1}$$

where $\mathbf{y}_d(t) \in \mathcal{R}^{n_{y_d}}$ denotes the generalized state variables, $\mathbf{u}(t) \in \mathcal{R}^{n_u}$ is a control function, by $\mathbf{p} \in \mathcal{R}^{n_p}$ the parameters constant in the time have been denoted. The independent variable $t \in [t_0 \quad t_F]$, in many application has a known physical interpretation like time or a length of e tubular reactor. Then, the vector-valued function F_d such, that is under considerations.

$$F_d : \mathcal{R}^{n_{y_d}} \times \mathcal{R}^{n_u} \times \mathcal{R}^{n_p} \times \mathcal{R} \rightarrow \mathcal{R}^{n_{y_d}} \tag{2}$$

In the descriptor processes, two main types of the generalized state variables have been specified. There are differential state variables $\mathbf{y}(t) \in \mathcal{R}^{n_y}$, as well as algebraic state variables $\mathbf{z}(t) \in \mathcal{R}^{n_z}$. Therefore, the vector of descriptor variables takes a particular form

$$\mathbf{y}_d(t) = \begin{bmatrix} \mathbf{y}(t) \\ \mathbf{z}(t) \end{bmatrix}. \tag{3}$$

Moreover, in the descriptor process a differential, as well as an algebraic part of equations system can be indicated. This specific representation is known as a *semi-explicit* form of the descriptor model

$$\begin{aligned} B_D\dot{\mathbf{y}}(t) &= F(\mathbf{y}(t), \mathbf{z}(t), \mathbf{u}(t), \mathbf{p}, t) \\ 0 &= G(\mathbf{y}(t), \mathbf{z}(t), \mathbf{u}(t), \mathbf{p}, t), \end{aligned} \tag{4}$$

where

$$F : \mathcal{R}^{n_y} \times \mathcal{R}^{n_z} \times \mathcal{R}^{n_u} \times \mathcal{R}^{n_p} \times \mathcal{R} \rightarrow \mathcal{R}^{n_y}, \tag{5}$$

and

$$G : \mathcal{R}^{n_y} \times \mathcal{R}^{n_z} \times \mathcal{R}^{n_u} \times \mathcal{R}^{n_p} \times \mathcal{R} \rightarrow \mathcal{R}^{n_z}. \tag{6}$$

Assumption 1. Let us assume, that

$$\det B_D \neq 0. \tag{7}$$

Then, the optimal control problem is to find the optimal control function $\mathbf{u}^\star(t)$, which minimizes the following process performance index

$$\min_{\mathbf{u}^\star(t)} \int_0^{t_F} \mathcal{L}(\mathbf{y}_d(t), \mathbf{u}(t), \mathbf{p}, t)dt + \mathcal{E}(\mathbf{y}_d(t_F)), \tag{8}$$

where

$$\mathcal{L} : \mathcal{R}^{n_{y_d}} \times \mathcal{R}^{n_u} \times \mathcal{R}^{n_p} \times \mathcal{R} \rightarrow \mathcal{R}, \tag{9}$$

and

$$\mathcal{E} : \mathcal{R}^{n_{y_d}} \rightarrow \mathcal{R}. \tag{10}$$

In the next section consistent initial conditions of the descriptor system have been discussed. After that, the descriptor system will be extended by the variability constraints.

3 Consistent Initial Conditions

Let us consider the algebraic part of the descriptor process

$$0 = G(\mathbf{y}(t), \mathbf{z}(t), \mathbf{u}(t), t). \tag{11}$$

The consistency of the model initial conditions can be easily checked by the Eq. (11). Moreover, this step does not require any other complicated calculations. The importance of this step has been highlighted by the Conditions 1 and 2.

Condition 1 *For a given control function* $\mathbf{u}(t)$ *and at given time* $t = t_0$, *the initial conditions consistency for the descriptor model* (4) *can be checked by the following equation*

$$0 = G(\mathbf{y}(t_0), \mathbf{z}(t_0), \mathbf{u}(t_0), t_0). \tag{12}$$

Because the initial conditions of the algebraic state variables are often unknown, therefore the Conditions 2 can be applied.

Condition 2 *The model equation* (11) *can be used to define the algebraic state variables* $\mathbf{z}(t)$ *as a function of the differential state variables* $\mathbf{y}(t)$, *given control function* $\mathbf{u}(t)$ *and the independent variable* t.

The local fulfillment of this two conditions can be checked by extension of function $G(\cdot)$ in Taylor series around a given point $(\mathbf{y}(t), \mathbf{z}(t)) = (\mathbf{x_y}, \mathbf{x_z})$ for $\mathbf{u}(t) = \mathbf{u}(t_0)$ and $t = t_0$

$$G(\mathbf{x_y} + \Delta\mathbf{x_y}, \mathbf{x_z} + \Delta\mathbf{x_z}, \mathbf{u}(t_0), t_0)$$

$$\approx \frac{\partial G}{\partial \mathbf{x_y}}\Delta\mathbf{x_y} + \frac{\partial G}{\partial \mathbf{x_z}}\Delta\mathbf{x_z} + G(\mathbf{x_y}, \mathbf{x_z}, \mathbf{u}(t_0), t_0) \tag{13}$$

$$= 0,$$

which results in a system of linear equations

$$\begin{bmatrix} G_{\mathbf{x_y}} & G_{\mathbf{x_z}} \end{bmatrix} \begin{bmatrix} \Delta\mathbf{x_y} \\ \Delta\mathbf{x_z} \end{bmatrix} = -G(\mathbf{x_y}, \mathbf{x_z}, \mathbf{u}(t_0), t_0) \tag{14}$$

and

$$G_{\mathbf{x_z}}\Delta\mathbf{x_z} = -G(\mathbf{x_y}, \mathbf{x_z}, \mathbf{u}(t_0), t_0)\Delta\mathbf{x_z} - G_{\mathbf{x_y}}\Delta\mathbf{x_y}. \tag{15}$$

Theorem 1 *For a given values $x_y = y(t_0)$ Eq. (15) has an unique solution if and only if, when*

$$\det G_{x_z} \neq 0. \tag{16}$$

It means, that for the given initial values of the differential state variables $y(t_0)$, the control function $u(t_0)$ and for a given time $t = t_0$, then the algebraic state variables can be well-defined.

4 Variability of the Differential State Variables

In this section, the differential part of the descriptor model (4) has been considered. The differential equations from the *semi-explicit* descriptor model takes a following form

$$B_D \dot{y}(t) = F(y(t), z(t), u(t), t), \tag{17}$$

with $rank\, B_D = n_y$, and

$$F : \mathcal{R}^{n_y} \times \mathcal{R}^{n_z} \times \mathcal{R}^{n_u} \times \mathcal{R} \to \mathcal{R}^{n_y}. \tag{18}$$

Let us assume, that $B_D = I$ and rank $I = n_y$, and consider both left and right hand-side of this equation extended in a Taylor series

$$\begin{aligned}
\dot{y} + \Delta \dot{y} &= F(y + \Delta y, z + \Delta z, u(t), t) \\
&= F(y, z, u(t), t) + F_y \Delta y + F_z \Delta z.
\end{aligned} \tag{19}$$

Let at the time $t = t_0$ the variability of the state variables is defined as

$$\dot{y}(t_0) = x_{\dot{y}_0}, \tag{20}$$

with $n_{\dot{y}} = n_y = n_{x_{\dot{y}_0}}$. Moreover, for a given time $t = t_0$, the control function $u(t_0)$ and a vector of initial conditions $\begin{bmatrix} x_{y_0} \\ x_{z_0} \end{bmatrix}$, the right hand-side equation of the system (17) is equal to

$$F(x_{y_0}, x_{z_0}, u(t_0), t_0). \tag{21}$$

Theorem 2 *At a given time $t = t_0$ and for a given control function $u(t_0)$, Eq. (17) is satisfied if and only if, when*

$$x_{\dot{y}_0} = F(x_{y_0}, x_{z_0}, u(t_0), t_0). \tag{22}$$

Proof Let $t = t_0$ and $\mathbf{u}(t_0)$ are known. Moreover, let $\mathbf{y}(t_0) = \mathbf{x}_{\mathbf{y}_0}$ and $\mathbf{z}(t_0) = \mathbf{x}_{\mathbf{z}_0}$.
Then

$$\dot{\mathbf{y}}(t) = F(\mathbf{y}(t), \mathbf{z}(t), \mathbf{u}(t), t)$$

$$\overset{t=t_0}{=} F(\mathbf{y}(t_0), \mathbf{z}(t_0), \mathbf{u}(t_0), t_0)$$

$$= F(\mathbf{x}_{\mathbf{y}(t_0)}, \mathbf{x}_{\mathbf{z}(t_0)}, \mathbf{u}(t_0), t_0) \tag{23}$$

$$= \mathbf{x}_{\dot{\mathbf{y}}_0},$$

and this should be proven.

Let us extend the left hand-side of the system (17) in the Taylor series

$$F_L(\dot{\mathbf{y}} + \Delta\dot{\mathbf{y}}) = F_L(\dot{\mathbf{y}}) + \frac{\partial F_L}{\partial \dot{\mathbf{y}}} \Delta\dot{\mathbf{y}}$$

$$= \dot{\mathbf{y}} + B_D \Delta\dot{\mathbf{y}} \tag{24}$$

$$= \mathbf{x}_{\dot{\mathbf{y}}} + B_D \Delta\mathbf{x}_{\dot{\mathbf{y}}}.$$

Let us perform the same with the right hand-side of the Eq. (17)

$$F_R(\mathbf{y}(t) + \Delta\mathbf{y}(t), \mathbf{z}(t) + \Delta\mathbf{z}(t))$$

$$= F_R(\mathbf{y}(t), \mathbf{z}(t)) + \frac{\partial F_R}{\partial \mathbf{y}} \Delta\mathbf{y} + \frac{\partial F_R}{\partial \mathbf{z}} \Delta\mathbf{z} \tag{25}$$

$$= F(\mathbf{x}_\mathbf{y}, \mathbf{x}_\mathbf{z}) + \frac{\partial F}{\partial \mathbf{x}_\mathbf{y}} \Delta\mathbf{x}_\mathbf{y} + \frac{\partial F}{\partial \mathbf{x}_\mathbf{z}} \Delta\mathbf{x}_\mathbf{z}.$$

According to the Theorem 2, Eqs. (24), (25) can be compared to each other

$$\mathbf{x}_{\dot{\mathbf{y}}} + B_D \Delta\mathbf{x}_{\dot{\mathbf{y}}} = F(\mathbf{x}_\mathbf{y}, \mathbf{x}_\mathbf{z}) + \frac{\partial F}{\partial \mathbf{x}_\mathbf{y}} \Delta\mathbf{x}_\mathbf{y} + \frac{\partial F}{\partial \mathbf{x}_\mathbf{z}} \Delta\mathbf{x}_\mathbf{z} \tag{26}$$

and presented as a system of linear equations in the following matrix form

$$\begin{bmatrix} B_D & F_\mathbf{y} & F_\mathbf{z} \end{bmatrix} \begin{bmatrix} \Delta\mathbf{x}_{\dot{\mathbf{y}}} \\ \Delta\mathbf{x}_\mathbf{y} \\ \Delta\mathbf{x}_{\dot{\mathbf{z}}} \end{bmatrix} = F(\mathbf{x}_\mathbf{y}, \mathbf{x}_\mathbf{z}, \mathbf{u}(t_0), t_0) - \mathbf{x}_{\dot{\mathbf{y}}}. \tag{27}$$

The relation between generalized state variables, which is indicated by the system
(27), was obtained by an analysis of the differential part of the descriptor model (4).

The local approximation of both differential and algebraic parts of the descriptor
model (4) resulted in a new system of linear equations in the following form

$$
\begin{bmatrix} F_{\dot{y}} & F_y & F_z \\ 0 & G_y & G_z \end{bmatrix} \begin{bmatrix} \Delta x_{\dot{y}} \\ \Delta x_y \\ \Delta x_{\dot{z}} \end{bmatrix} = \begin{bmatrix} F(x_y, x_z, u(t_0), t_0) - x_{\dot{y}} \\ G(x_y, x_z, u(t_0), t_0) \end{bmatrix}. \tag{28}
$$

Because in the system (28) the number of unknown is equal to $n_y + n_y + n_z$ and

$$
rank \begin{bmatrix} F_{\dot{y}} & F_y & F_z \\ 0 & G_y & G_z \end{bmatrix} = n_y + n_z, \tag{29}
$$

then the system has infinitely many solutions. Moreover, according to the Kronecker-Capelly theorem, the solution of the system (28) is dependent on n_y parameters.

Before the system (28) will be solved, let us consider a homogeneous system

$$
\begin{bmatrix} F_{\dot{y}} & F_y & F_z \\ 0 & G_y & G_z \end{bmatrix} \begin{bmatrix} \Delta x_{\dot{y}} \\ \Delta x_y \\ \Delta x_z \end{bmatrix} = \begin{bmatrix} 0 \\ 0 \end{bmatrix}. \tag{30}
$$

and let us find its null-solution.

In order to solve the system (30), let us write it in a new, but equivalent form

$$
\begin{bmatrix} G_z & G_y & 0 \\ F_z & F_y & F_{\dot{y}} \end{bmatrix} \begin{bmatrix} \Delta x_z \\ \Delta x_y \\ \Delta x_{\dot{y}} \end{bmatrix} = \begin{bmatrix} 0 \\ 0 \end{bmatrix}. \tag{31}
$$

The form (31) enables us to solve the system (30) with respect to the variables Δx_z and Δx_y, as well as to find the null space $\mathcal{N}$ of the system (30). Therefore, let us obtain the (*reduced row echelon form*) of the considered matrix.

If $\det G_z \neq 0$, then the reduced form of the system (31) can be obtained using four linear transformations P_1, P_2, P_3 and P_4.

1. The transformation P_1 denotes a such linear rows combination of the matrix G_z, that enables us to transform F_z matrix into the null matrix

$$
P_1 G_z + F_z = 0, \tag{32}
$$

then

$$
P_1 = -F_z G_z^{-1}, \tag{33}
$$

and it results in

$$
\begin{bmatrix}
G_z & G_y & 0 \\
0 & -F_z G_z^{-1} G_y + F_y & F_{\dot{y}}
\end{bmatrix}.
\tag{34}
$$

2. The transformation P_2 denotes a such linear rows combination of the matrix $(-F_z G_z^{-1} G_y + F_y)$, that enables us to transform G_y into the null matrix

$$
P_2(-F_z G_z^{-1} G_y + F_y) + G_y = 0,
\tag{35}
$$

then

$$
P_2 = -G_y(-F_z G_z^{-1} G_y + F_y)^{-1},
\tag{36}
$$

and it results in

$$
\begin{bmatrix}
G_z & 0 & -G_y(-F_z G_z^{-1} G_y + F_y)^{-1} F_{\dot{y}} \\
0 & -F_z G_z^{-1} G_y + F_y & F_{\dot{y}}
\end{bmatrix}.
\tag{37}
$$

3. The transformation P_3 makes from the matrix G_z the identity matrix

$$
P_3 G_z = 1,
\tag{38}
$$

then

$$
P_3 = G_z^{-1}
\tag{39}
$$

and it results in

$$
\begin{bmatrix}
1 & 0 & -G_z^{-1} G_y(-F_z G_z^{-1} G_y + F_y)^{-1} F_{\dot{y}} \\
0 & -F_z G_z^{-1} G_y + F_y & F_{\dot{y}}
\end{bmatrix}.
\tag{40}
$$

4. The transformation P_4 makes from the matrix $(-F_z G_z^{-1} G_y + F_y)$ the identity matrix

$$
P_4(-F_z G_z^{-1} G_y + F_y) = 1,
\tag{41}
$$

then

$$
P_4 = (-F_z G_z^{-1} G_y + F_y)^{-1}
\tag{42}
$$

and it results in

$$
\begin{bmatrix}
1 & 0 & -G_z^{-1} G_y(-F_z G_z^{-1} G_y + F_y)^{-1} F_{\dot{y}} \\
0 & 1 & (-F_z G_z^{-1} G_y + F_y)^{-1} F_{\dot{y}}
\end{bmatrix}.
\tag{43}
$$

The system (43) indicates the necessary conditions of the problem solvability

Theorem 3 *The necessary solvability condition of the system (43) is*

$$\det(-F_z G_z^{-1} G_y + F_y) \neq 0, \tag{44}$$

which denotes the linear independence of the functions F and G according to the unknown parameters Δx_z and Δx_y at a given point

$$\begin{bmatrix} x_{y_0} \\ x_{z_0} \end{bmatrix} = \begin{bmatrix} y(t_0) \\ z(t_0) \end{bmatrix}. \tag{45}$$

The null solution, which can be obtained for a given value of a derivative $x_{\dot{y}}$ at the point $[x_{y_0} \; x_{z_0}]^T$, takes the following form

$$\begin{bmatrix} \Delta x_z \\ \Delta x_y \\ \Delta x_{\dot{y}} \end{bmatrix} = \begin{bmatrix} G_z^{-1} G_y (-F_z G_z^{-1} G_y + F_y)^{-1} F_{\dot{y}} c \\ -(-F_z G_z^{-1} G_y + F_y)^{-1} F_{\dot{y}} c \\ c \end{bmatrix}, \tag{46}$$

where $c \in \mathcal{R}^{n_{\dot{y}}}$ can take a given value.

As it can be observed, the null space can be defined as

$$\mathcal{N}(A) = span \left\{ \begin{bmatrix} G_z^{-1} G_y (-F_z G_z^{-1} G_y + F_y)^{-1} F_{\dot{y}} c \\ -(-F_z G_z^{-1} G_y + F_y)^{-1} F_{\dot{y}} c \\ c \end{bmatrix} \right\}. \tag{47}$$

The next question is the form of a particular solution. Let us consider the system of linear equations, which has been obtained by Taylor expansion of the differential-algebraic system (28). The systems (28) after the transformation $P_1 = -F_z G_z^{-1}$ has been taken a form

$$\begin{bmatrix} G_z & G_y & 0 \\ 0 & -F_z G_z^{-1} G_y + F_y & F_{\dot{y}} \end{bmatrix} \begin{bmatrix} \Delta x_z \\ \Delta x_y \\ \Delta x_{\dot{y}} \end{bmatrix} = \begin{bmatrix} G(x_y, x_z) \\ F(x_{\dot{y}}, x_y, x_z) - F_z G_z^{-1} G \end{bmatrix}. \tag{48}$$

Let $\Delta x_{\dot{y}} \equiv 0$, then

$$(-F_z G_z^{-1} G_y + F_y) \Delta x_y = -F(x_{\dot{y}}, x_y, x_z) + F_z G_z^{-1} G. \tag{49}$$

If $\det(-F_{\mathbf{z}}G_{\mathbf{z}}^{-1}G_{\mathbf{y}} + F_{\mathbf{y}}) \neq 0$, then

$$\Delta\mathbf{x_y} = (-F_{\mathbf{z}}G_{\mathbf{z}}^{-1}G_{\mathbf{y}} + F_{\mathbf{y}})^{-1}(-F(\mathbf{x_{\dot{y}}}, \mathbf{x_y}, \mathbf{x_z}) + F_{\mathbf{z}}G_{\mathbf{z}}^{-1}G). \tag{50}$$

The expression for $\Delta\mathbf{x_z}$ can be obtained by the first relation in the system of linear equations

$$G_{\mathbf{z}}\Delta\mathbf{x_z} + G_{\mathbf{y}}\Delta\mathbf{x_y} = -G(\mathbf{x_y}, \mathbf{x_z}), \tag{51}$$

and

$$\Delta\mathbf{x_z} = -G_{\mathbf{z}}^{-1}(G(\mathbf{x_y}, \mathbf{x_z}) + G_{\mathbf{y}}\Delta\mathbf{x_y}), \tag{52}$$

which can obtained if and only if $G_{\mathbf{z}}^{-1}$ exists and $\Delta\mathbf{x_y}$ is defined by the Eq. (50).

The presented considerations result in some important consequences

1. A new type of additional restrictions, which can be considered in the descriptor processes, have been introduced. *The variability constraints* can be imposed on the rate of changes.
2. The restrictions on the rate of changes can have important practical applications, because they reflect the safety principles. The variability constraints can represent the maximum allowable rate of changes of the aircraft's flight parameters.
3. The constraints on the state variability define also the consistent initial conditions of the generalized state variables

$$\begin{bmatrix} \mathbf{z}(t_0) \\ \mathbf{y}(t_0) \\ \dot{\mathbf{y}}(t_0) \end{bmatrix} = \begin{bmatrix} \mathbf{x_{z_0}} \\ \mathbf{x_{y_0}} \\ \mathbf{x_{\dot{y}_0}} \end{bmatrix} = \begin{bmatrix} \mathbf{x_z} \\ \mathbf{x_y} \\ \mathbf{x_{\dot{y}}} \end{bmatrix} + \begin{bmatrix} \Delta\mathbf{x_z} \\ \Delta\mathbf{x_y} \\ \Delta\mathbf{x_{\dot{y}}} \end{bmatrix}, \tag{53}$$

where

$$\begin{bmatrix} \Delta\mathbf{x_z} \\ \Delta\mathbf{x_y} \\ \Delta\mathbf{x_{\dot{y}}} \end{bmatrix} = \begin{bmatrix} -G_{\mathbf{z}}^{-1}(G(\mathbf{x_y}, \mathbf{x_z}) + G_{\mathbf{y}}\Delta\mathbf{x_y}) \\ (-F_{\mathbf{z}}G_{\mathbf{z}}^{-1}G_{\mathbf{y}} + F_{\mathbf{y}})^{-1}(-F(\mathbf{x_{\dot{y}}}, \mathbf{x_y}, \mathbf{x_z}) + F_{\mathbf{z}}G_{\mathbf{z}}^{-1}G) \\ \mathbf{0} \end{bmatrix} + $$

$$\mathbf{c} \begin{bmatrix} G_{\mathbf{z}}^{-1}G_{\mathbf{y}}(-F_{\mathbf{z}}G_{\mathbf{z}}^{-1}G_{\mathbf{y}} + F_{\mathbf{y}})^{-1}F_{\dot{\mathbf{y}}} \\ -(-F_{\mathbf{z}}G_{\mathbf{z}}^{-1}G_{\mathbf{y}} + F_{\mathbf{y}})^{-1}F_{\dot{\mathbf{y}}} \\ \mathbf{1} \end{bmatrix}. \tag{54}$$

5 Summary

The main result of the presented work is a new method to find the consistent initial conditions for solving the DAE systems. To find the initial values of the descriptor variables, the variability constraints have been introduced.

Then, the control of descriptor systems with the variability constraints has been discussed. The presented considerations can be applied in the vision-based monitoring systems.

The presented method can answer the important question, how quickly can the state be changed, to preserve a failure-free process flow?

The future work will be concerned on the variability constraints applications in various branches of environmental engineering and technology.

Acknowledgements This work has been supported by the National Science Center under grant: DEC-2012/07/B/ST7/01216.

References

1. An, Y.-K., Yang, J., Hwang, S., Sohn, H.: Line laser lock-in thermography for instantaneous imaging of cracks in semiconductor chips. Opt. Lasers Eng. **73**, 128–136 (2015). doi:10.1016/j.optlaseng.2015.04.013
2. Aubry-Wake, C., Baraer, M., McKenzie, J.M., Mark, B.G., Wigmore, O., Hellström, R.A., Lautz, L., Somers, L.: Measuring glacier surface temperatures with ground-based thermal infrared imaging. Geophys. Res. Lett. **42**, 8489–8497 (2015). doi:10.1002/2015GL065321
3. Betts, J.T.: Practical Methods for Optimal Control and Estimation Using Nonlinear Programming, 2nd edn. SIAM, Philadelphia (2010). doi:10.1137/1.9780898718577.
4. Biegler, L.T.: Nonlinear programming strategies for dynamic chemical process optimization. Theor. Found. Chem. Eng. **48**, 541–554 (2014). doi:10.1134/S0040579514050157
5. Biegler, L.T., Campbell, S.L., Mehrmann, V. (eds.): Control and optimization with differential-algebraic constraints. Society for Industrial and Applied Mathematics (2012). doi:10.1137/9781611972252.fm
6. Brenan, K.E., Campbell, S.L., Petzold, L.R.: Numerical Solution of Initial-Value Problems in Differential Algebraic Equations. SIAM, Philadelphia (1996). doi:10.1137/1.9781611971224
7. Petzold, L.: Differential/algebraic equations are not ODEs. SIAM J. Sci. Stat. Comput. **3**, 367–384 (1982). doi:10.1137/0903023

Comparison of Different ACO Start Strategies Based on InterCriteria Analysis

Olympia Roeva, Stefka Fidanova and Marcin Paprzycki

Abstract In the combinatorial optimization, the goal is to find the optimal object from a finite set of objects. From computational point of view the combinatorial optimization problems are hard to be solved. Therefore on this kind of problems usually is applied some metaheuristics. One of the most successful techniques for a lot of problem classes is metaheuristic algorithm Ant Colony Optimization (ACO). Some start strategies can be applied on ACO algorithms to improve the algorithm performance. We propose several start strategies when an ant chose first node, from which to start to create a solution. Some of the strategies are base on forbidding some of the possible starting nodes, for one or more iterations, because we suppose that no good solution starting from these nodes. The aim of other strategies are to increase the probability to start from nodes with expectations that there are good solutions starting from these nodes. We can apply any of the proposed strategy separately or to combine them. In this investigation InterCriteria Analysis (ICrA) is applied on ACO algorithms with the suggested different start strategies. On the basis of ICrA the ACO performance is examined and analysed.

Keywords InterCriteria analysis · Ant colony optimization · Start strategies · Multiple knapsack problem

O. Roeva
Institute of Biophysics and Biomedical Engineering, Bulgarian Academy of Sciences,
Sofia, Bulgaria
e-mail: olympia@biomed.bas.bg

S. Fidanova (✉)
Institute of Information and Communication Technology, Bulgarian Academy of Sciences,
Sofia, Bulgaria
e-mail: stefka@parallel.bas.bg

M. Paprzycki
System Research Institute, Polish Academy of Sciences Warsaw and Management Academy,
Warsaw, Poland
e-mail: marcin.paprzycki@ibspan.waw.pl

© Springer International Publishing AG 2018
S. Fidanova (ed.), *Recent Advances in Computational Optimization*,
Studies in Computational Intelligence 717, DOI 10.1007/978-3-319-59861-1_4

1 Introduction

Many real-world problems can be describe as combinatorial optimization problems. We can mention Traveling Salesman Problem [22], Vehicle Routing [24], Minimal Spanning Tree [18], Constrain Satisfaction [16], Knapsack Problem [12] and many others. They are NP-hard problems and spends a lot of computational resources. Therefore to find close to optimal solution for a reasonable time, metaheuristic methods are applied.

Ant Colony Optimization (ACO) is between of the very successful metaheuristics [14]. The idea for ACO comes from real ants behaviour and more specifically, the way they look for a food. The ACO algorithm is proposed by Marco Dorigo more than 20 years ago [11]. Later several variants and supplements are added to the algorithm to improve its performance [10]. In [13] various ACO start strategies, which lead to finding better solutions, are proposed.

InterCriteria Analysis (ICrA) is an approach aiming to go beyond the nature of the criteria involved in a process of evaluation of multiple objects against multiple criteria, and, thus to discover some dependencies between the ICrA criteria themselves [4]. For the first time ICrA has been applied for the purposes of temporal, threshold and trends analyses of an economic case-study of European Union member states' competitiveness [5–7]. Further, ICrA has been used to discover the considered dependencies in a lot of problems, for example for the in-depth analysis of deceases [23]. Another example is the ICrA application for investigation of correlations of the parameters in various mathematical models and performance criteria for metaheuristics as GAs and ACO [1, 19, 20].

In this paper ICrA is applied for analysis of an ACO algorithm with various start strategies. ACO is applied to find near-optimal solutions for Multiple Knapsack Problem (MKP). The aim is to analyze the algorithm performance according the start strategies and to study the correlations between the strategies.

The organization of the rest of the paper is as follows. In Sect. 2 the ACO algorithm with start strategies is described. In Sect. 3 the problem on which we test the strategies influence is presented. Section 4 is dedicated to ICrA and in Sect. 5 are studied the relations between the start strategies. The conclusions is given in Sect. 6.

2 ACO Algorithm with Start Strategies

The idea for ACO algorithm comes from the observation of real ant behaviour when they look for a food [10, 11]. The solved problem is represented by graph called graph of the problem. The feasible solutions are represented by path in a graph. When we solve the problem we look for a shorter path (if it is minimization problem) or longer path (if it is maximization problem). The ACO is an constructive method and it not needs an initial solution. In every iteration the ant starts from random node and creates the solution. Random start is a way of search diversification in the search space. If the last selected node is u the ant selects the next node (v) to be included in the decision applying probabilistic rule called transition probability.

$$
p_{uv} = \tau_{uv}^{\alpha}\eta_{uv}^{\beta} \left[\sum_{(u,w)\in E_S \,:\, w\not\subset X} \left(\tau_{uw}^{\alpha}\eta_{uw}^{\beta}\right)^{-1} \right], \tag{1}
$$

where α and β are transition probability parameters. At the beginning the value of the pheromone on all elements is the same and is set to a value between 0 and 1. In each iteration we update the pheromone on the elements of the graph, according to the value of the objective function. The elements belonging to the better solutions receive more pheromone then others. The pheromone shows the global memory of the ants. The better paths (solutions) accumulate more pheromone than others. It is kind of intensification of the search around good found solutions.

The random start is very important for good performance of the ACO algorithm, but for some classes of problems the mode of choosing starting node can be significant. Between them are subset problems. For better managing the search process we include semi-random start of the ants. Our aim is to use the ants experience to solve the problem in more efficient way. The set of nodes is divided into several subsets. An estimation of how good and how bad is to start from some subset is introduced according the number of good and bed solutions started from this subset.

$$
D_j(i) = \phi.D_j(i-1) + (\psi-\phi).F_j(i), \tag{2}
$$

$$
E_j(i) = \phi.E_j(i-1) + (\psi-\phi).G_j(i), \tag{3}
$$

where $i \geq 1$ is the current process iteration and for each j ($1 \leq j \leq N$):

$$
F_j(i) = \begin{cases} f_{j,A}/n_j & \text{if } n_j \neq 0 \\ \\ F_j(i-1) & \text{otherwise} \end{cases}, \tag{4}
$$

$$
G_j(i) = \begin{cases} g_{j,B}/n_j & \text{if } n_j \neq 0 \\ \\ G_j(i-1) & \text{otherwise} \end{cases}, \tag{5}
$$

$f_{j,A}$ is the number of the solutions among the best $A\%$, $g_{j,B}$ is the number of the solutions among the worst $B\%$, where $A + B \leq 100$, $i \geq 2$ and

$$
\sum_{j=1}^{N} n_j = n, \tag{6}
$$

where n_j ($1 \leq j \leq N$) is the number of solutions obtained by ants starting from nodes subset j, n is the number of ants. Initial values of the weight coefficients are:

$D_j(1) = 1$ and $E_j(1) = 0$. The parameters ϕ and ψ, $0 \leq \phi$, $\psi \leq 1$ and $\psi \geq \phi$, show the weight of the information from the previous iterations and from the last iteration. When $\phi = 0$ only the information from the last iteration is taken in to account. If $\phi = 0.5 \times \psi$ the influence of the previous iterations versus the last is equal. When $\phi = \psi$ only the information from the previous iterations is taken in to account. When $\phi = 0.25 \times \psi$ the weight of the information from the previous iterations is three times less than this one of the last iteration. When $\phi = 0.75 \times \psi$ the weight of the previous iterations is three times higher than this one of the last iteration. This kind of estimation, where the sum of the weight coefficients is less or equal to 1 is called intuitionistic fuzzy estimation [3]. When $\psi = 1$ the estimation is fuzzy.

Several start strategies and combinations of them are proposed. For every subset j, $D_j(i)$ is the estimation how good is to start from the subset j and $E_j(i)$ is the estimation how bad is to start from the subset j, where i is the iteration number. Than a thresholds for good estimation D and for bad estimation E are fixed. The proposed start strategies are as follows [13]:

(1) If $\frac{E_j(i)}{D_j(i)} > E$ then for current iteration the subset j is forbidden. The starting node is randomly chosen from $\{j \mid j$ is not forbidden$\}$;
(2) If $\frac{E_j(i)}{D_j(i)} > E$ then for current simulation the subset j is forbidden. The starting node is randomly chosen from $\{j \mid j$ is not forbidden$\}$;
(3) If $\frac{E_j(i)}{D_j(i)} > E$ then for K_1 consecutive iterations the subset j is forbidden. The starting node is randomly chosen from $\{j \mid j$ is not forbidden$\}$;
(4) Let $r_1 \in [\frac{1}{2}, 1)$ and $r_2 \in [0, 1]$ to be random numbers. If $r_2 > r_1$ a node from subset $\{j \mid D_j(i) > D\}$ is randomly chosen, otherwise a node from the not forbidden subsets is randomly chosen. r_1 is chosen and fixed at the beginning.
(5) Let $r_1 \in [\frac{1}{2}, 1)$ and $r_2 \in [0, 1]$ to be random numbers. If $r_2 > r_1$ a node from subset $\{j \mid D_j(i) > D\}$ is randomly chosen, otherwise a node from the not forbidden subsets is randomly chosen. r_1 is chosen at the beginning and increase with r_3 every iteration.

Here K_1, $K_1 \in [0, \text{number of iterations}]$ is a parameter.

We can apply one of the start strategies or to combine some of them. The Strategies 1, 2, and 3 can be combined with Strategies 4 and 5. When an ant chooses a start node first is applied Strategy 1, 2, or 3 and after that Strategy 5 or 6. Thus together with completely random start there are 12 strategies. More details about ACO with semi random start can be seen in [13].

3   Multiple Knapsack Problem

The start node selection is very important for subset problems. On this kind of problems only part of the nodes of the graph of the problem belong to the solution. Thus if the start node do not belong to any good solution, for the ant is impossible to construct close to optimal solution. The Multiple Knapsack Problem (MKP) is a representative of the class of subset problems. A lot of problems can be defined as MKP.

It also arise as a sub-problem in a group of more complex problems and these algorithms will benefit from any improvement algorithm for solving MKP. Some of important applications that can be formulated as MKP are cargo loading problems, cutting stock, bin-packing, budget control and financial management. The MKP is used in fault tolerance problem [21]. Authors in [9] designed a public cryptography scheme whose security realize on the difficulty of solving the MKP. In [17] two-processor scheduling problems are proposed to be solved as a MKP. We will mention other applications as industrial management, naval, aerospace, computational complexity theory, etc.

Sports management deals with the problem of optimal location of limited resources (money) to meet given objects (championships). The existing goal of any team's managements to find the subset of players capable of playing in the important gain without exceeding the limited budget. In a transportation network the shortest path problem determines the subset of the connected roads that collectively comprise (i) the shortest driving distance, (ii) the smallest driving time or (iii) the cheapest fair between two cities. The problem is what subset of lines gives the faster response time for communication between them. Complexity theory is the part of the theory of computation of the resources required to solve a given problem.

Where a general problems is transformed to a MKP or a MKP appears as a sub-problem the more theoretical applications are appeared. When solving the generalized assignment problem (vehicle routing problem) MKP appears as a sub-problem. Moreover, MKP can be presented as a general model for binary problem with positive coefficients [15].

The MKP can be formulated as follows:

$$\max \sum_{j=1}^{n} p_j x_j$$

$$\text{subject to } \sum_{j=1}^{n} r_{ij} x_j \leq c_i \quad i = 1, \ldots, m \tag{7}$$

$$x_j \in \{0, 1\} \quad j = 1, \ldots, n$$

$$x_j = \begin{cases} 1 \text{ iff the object } j \text{ is chosen,} \\ 0 \text{ otherwise.} \end{cases}$$

where m are the resources (the knapsacks), n are the objects, p_j is a profit of every object j, c_j (knapsack capacity) is resource budget, r_{ij} is the consumption of resource i by object j.

The goal is maximizing the sum of the profits for a limited budget.

There are m constraints in this problem, so MKP is also called m-dimensional knapsack problem. Let

$$I = \{1, \ldots, m\}, \, J = \{1, \ldots, n\},$$

with $c_i \geq 0$ for all $i \in I$.

A well-stated MKP assumes that

$$p_j > 0, \; r_{ij} \leq c_i \leq \sum_{j=1}^{n} r_{ij}$$

for all

$$i \in I, \; j \in J.$$

Note that the $[r_{ij}]_{m \times n}$ matrix and $[c_i]_m$ vector are both non-negative.

One of the basic elements of the ACO metaheuristic is the representation of the problem by graph, thus a path through the graph represents a solution to the problem. The graph of the problem is defined as follows: the nodes correspond to the items, the arcs fully connect nodes. Fully connected graph means that after the object i one can chooses the object j for every i and j if there are enough resources and object j is not chosen yet. When walking through the graph, the ants deposit a pheromone on the arcs.

4 InterCriteria Analysis

Let us be given an index matrix (IM) [2] whose index sets for rows consist of the names of the criteria and for columns – objects. We will obtain an IM with index sets consisting of the names of the criteria both for rows and for columns. The elements intuitionistic fuzzy pairs (IFPs) of this IM corresponds to the degrees of "agreement" and degrees of "disagreement" of the considered criteria.

Further, by O we denote the set of all objects $O_1, O_2, \ldots, O_n$ being evaluated, and by $C(O)$ the set of values assigned by a given criteria C to the objects, i.e.

$$O \stackrel{\text{def}}{=} \{O_1, O_2, \ldots, O_n\},$$

$$C(O) \stackrel{\text{def}}{=} \{C(O_1), C(O_2), \ldots, C(O_n)\}.$$

Let $x_i = C(O_i)$. Then the following set can be defined:

$$C^*(O) \stackrel{\text{def}}{=} \{\langle x_i, x_j \rangle | i \neq j \, \& \, \langle x_i, x_j \rangle \in C(O) \times C(O)\}.$$

In order to find the degrees of "agreement" of two criteria the vector of all internal comparisons of each criteria is constructed. This vector fulfil exactly one of the following three relations – R, $\overline{R}$ and $\tilde{R}$. For a fixed criterion C and any ordered pair $\langle x, y \rangle \in C^*(O)$ it is required:

$$\langle x, y \rangle \in R \Leftrightarrow \langle y, x \rangle \in \overline{R} \tag{8}$$

$$\langle x, y \rangle \in \tilde{R} \Leftrightarrow \langle x, y \rangle \notin (R \cup \overline{R}) \tag{9}$$

$$R \cup \overline{R} \cup \tilde{R} = C^*(O) \tag{10}$$

From the above it is seen that We only need to consider a subset of $C(O) \times C(O)$ for the effective calculation of $V(C)$ (vector of internal comparisons). From Eqs. (8)–(10) it follows that if we know what is the relation between x and y, we also know what is the relation between y and x. Thus, we will only consider lexicographically ordered pairs $\langle x, y \rangle$.

Let:

$$C_{i,j} = \langle C(O_i), C(O_j) \rangle.$$

We construct the vector with exactly $\frac{n(n-1)}{2}$ elements:

$$V(C) = \{C_{1,2}, C_{1,3}, \ldots, C_{1,n}, C_{2,3}, C_{2,4}, \ldots, C_{2,n},$$

$$C_{3,4}, \ldots, C_{3,n}, \ldots, C_{n-1,n}\}.$$

for a fixed criterion C.

Further, we replace the vector $V(C)$ with $\hat{V}(C)$, where for each $1 \leq k \leq \frac{n(n-1)}{2}$ for the k-th component it is true:

$$\hat{V}_k(C) = \begin{cases} 1 & \text{iff } V_k(C) \in R, \\ -1 & \text{iff } V_k(C) \in \overline{R}, \\ 0 & \text{otherwise.} \end{cases}$$

We determine the degree of "agreement" ($\mu_{C,C'}$) between the two criteria as the number of matching components. The degree of "disagreement" ($\nu_{C,C'}$) is the number of components of opposing signs in the two vectors.

It is obvious that:

$$\mu_{C,C'} = \mu_{C',C},$$

$$\nu_{C,C'} = \nu_{C',C},$$

and $\langle \mu_{C,C'}, \nu_{C,C'} \rangle$ is an IFP.

The difference

$$\pi_{C,C'} = 1 - \mu_{C,C'} - \nu_{C,C'} \tag{11}$$

is considered as a degree of "uncertainty".

5 Numerical Results

We apply ICrA on the results achieved by ACO with different start strategies applied
on MKP [13]. 10 test problems from "OR-Library" (available within WWW access
at http://people.brunel.ac.uk/mastjjb/jeb/orlib) are used. The
problem is considered with 100 objects and 10 constraints. To provide a fair comparison for the above implemented ACO algorithm, a predefined number of iterations,
$k = 100$, is fixed for all the runs. The developed technique has been coded in C^{++}
language and implemented on a Pentium 4 (2.8 GHz). The parameters are fixed as
follows: $\rho = 0.5$, $a = 1$, $b = 1$, number of used ants is 20, the start area is divided
to 50 subsets, $A = 30$, $B = 30$, $D = 1$, $E = 0.1$, $K_1 = K_2 = 5$, $r_3 = 0.01$.

The ACO algorithm is performed with various number of the nodes in the nodes
subsets. The nodes subsets consist of the same number of nodes which vary from 1
to 10, namely 1, 2, 4, 5 and 10. The average results over the 10 test problems and 30
runs of the every problem with every one of the strategies are obtained. The ranking
is from 10 to 100. The obtained results are presented in Table 1.

From Table 1 it can be seen that the ICrA objects $(O_1, O_2, \ldots, O_{20})$ are the
different conditions, namely nodes 1, 2, 4, 5 and 10 in four cases of φ, $\varphi =$
[0 0.25 0.5 0.75]. The ICrA criteria $(C_1, C_2, \ldots, C_{12})$ are 12 different start strategies
for ACO. In the case of intuitionistic fuzzy estimation the ACO algorithm is again
performed with various number of the nodes. The results are presented in Table 2.

Analogically the nodes subsets consist of the same number of nodes which vary
from 1 to 10, namely 1, 2, 4, 5 and 10. The average results over the 10 test problems
and 30 runs of the every problem with every one of the strategies are obtained. The
results are ranked in the scale from 10 to 100. The value of φ varies from 0 to 0.75,
and the value of ψ varies from 0.25 to 0.825.

Analysing the obtained results the proposed in [8] consonance and dissonance
scale will be used. The scheme for defining the consonance and dissonance between
each pair of criteria is presented in Table 3.

After ICrA application we obtained the two IM with the relations between considered 12 criteria. The resulting IMs for $\mu_{C,C'}$ and $\nu_{C,C'}$ values are shown in Tables 4
and 5.

The obtained ICrA results are visualized on Fig. 1 within the specific triangular
geometrical interpretation of intuitionistic fuzzy sets, thus allowing us to order these
results according simultaneously to the degrees of "agreement" and "disagreement"
of the IFPs.

Figure 1 shows that there are some of the results with very high degree of "uncertainty". These results are due to the fact that for some of the strategies there are the
same results for the several ICrA objects. For example, see results for the Strategys
4 and 5 in the Table 1. Further, these results will not be discussed.

Table 1 Estimation of strategies

Criteria	Strategies	$\varphi = 0$					$\varphi = 0.25$				
		10	5	4	2	1	10	5	4	2	1
		O_1	O_2	O_3	O_4	O_5	O_6	O_7	O_8	O_9	O_{10}
C_1	Random	32	32	32	32	32	32	32	32	32	32
C_2	Strategy 1	84	84	87	83	83	83	88	86	90	90
C_3	Strategy 2	33	31	36	53	74	32	31	36	61	81
C_4	Strategy 3	79	86	86	88	86	62	86	84	84	96
C_5	Strategy 4	86	86	86	86	86	86	86	86	86	86
C_6	Strategy 5	86	86	86	86	86	86	86	86	86	86
C_7	Strategies 1–4	83	89	84	81	89	84	91	87	92	96
C_8	Strategies 1–5	83	89	84	81	89	84	91	87	92	96
C_9	Strategies 2–4	33	36	35	53	82	34	33	35	59	85
C_{10}	Strategies 2–5	33	36	35	63	82	34	33	35	59	85
C_{11}	Strategies 3–4	69	89	88	87	90	69	83	86	84	97
C_{12}	Strategies 3–5	69	89	88	87	90	69	83	86	84	97
Criteria	Strategies	$\varphi = 0.5$					$\varphi = 0.75$				
		10	5	4	2	1	10	5	4	2	1
		O_{11}	O_{12}	O_{13}	O_{14}	O_{15}	O_{16}	O_{17}	O_{18}	O_{19}	O_{20}
C_1	Random	32	32	32	32	32	32	32	32	32	32
C_2	Strategy 1	78	86	88	92	96	71	81	85	89	92
C_3	Strategy 2	34	35	38	51	78	35	55	52	60	87
C_4	Strategy 3	61	86	88	94	97	56	76	88	95	95
C_5	Strategy 4	86	86	86	86	86	86	86	86	86	86
C_6	Strategy 5	86	86	86	86	86	86	86	86	86	86
C_7	Strategies 1–4	79	90	87	94	97	67	83	89	94	95
C_8	Strategies 1–5	79	90	87	94	97	67	83	89	94	95
C_9	Strategies 2–4	35	40	44	56	83	39	47	48	58	85
C_{10}	Strategies 2–5	35	40	44	56	83	39	47	48	58	85
C_{11}	Strategies 3–4	68	92	88	92	96	56	81	87	94	97
C_{12}	Strategies 3–5	68	92	88	92	96	56	81	87	94	97

For better understanding of the results the values of the $\mu_{C,C'}$, $\nu_{C,C'}$, $\pi_{C,C'}$ of the criteria pairs are sorted by the value of the $\mu_{C,C'}$. The list is presented in Tables 6 and 7. Table 6 shows the criteria pair with high degrees of "agreement" ($\mu_{C,C'}$) and low value for the degree of "disagreement" ($\nu_{C,C'}$). Table 7 shows the criteria pair with high degree of "uncertainty".

Table 2 Intuitionistic fuzzy estimation of strategies

Criteria	Strategies	φ/ψ				
		0.125/0.25	0.125/0.5	0.125/0.75	0.125/0.875	0.25/0.5
		O_1	O_2	O_3	O_4	O_5
C_1	Random	32	32	32	32	32
C_2	Strategy 1	95	93	93	93	93
C_3	Strategy 2	82	79	79	79	79
C_4	Strategy 3	93	92	92	92	92
C_5	Strategy 4	83	83	83	83	83
C_6	Strategy 5	83	83	83	83	83
C_7	Strategies 1–4	96	96	96	96	96
C_8	Strategies 1–5	96	96	96	96	96
C_9	Strategies 2–4	84	83	83	83	83
C_{10}	Strategies 2–5	84	83	83	83	83
C_{11}	Strategies 3–4	94	93	93	93	93
C_{12}	Strategies 3–5	94	93	93	93	93

Criteria	Strategies	φ/ψ					
		0.25/0.75	0.25/0.875	0.5/0.75	0.5/0.875	0.75/0.875	0.5/1
		O_6	O_7	O_8	O_9	O_{10}	O_{11}
C_1	Random	32	32	32	32	32	32
C_2	Strategy 1	92	92	94	93	93	96
C_3	Strategy 2	85	85	82	77	83	78
C_4	Strategy 3	93	94	99	94	93	97
C_5	Strategy 4	83	83	83	83	83	83
C_6	Strategy 5	83	83	83	83	83	83
C_7	Strategies 1–4	96	96	92	94	95	97
C_8	Strategies 1–5	96	96	92	94	95	97
C_9	Strategies 2–4	83	83	82	81	86	83
C_{10}	Strategies 2–5	83	83	82	81	86	83
C_{11}	Strategies 3–4	94	94	93	93	97	96
C_{12}	Strategies 3–5	94	94	93	93	97	96

Regarding Tables 6 and 7 we observe that relations between criterion C_1 and criteria C_5, C_6 have the highest value of $\mu_{C,C'}$ ($\mu_{C,C'} = 1$), i.e., these criteria are in strong positive consonance. It means that the ACO algorithm performs in a similar way with random start and start Strategies 4 and 5.

In strategies 4 and 5 there are not forbidden regions similar to the random start. In these cases only the probability to choose the next element in the solution is different. The criteria pairs that also have the highest value of $\mu_{C,C'}$ ($\mu_{C,C'} = 1$) are $C_7 - C_8$ and $C_{11} - C_{12}$. These strategies (Strategies 1–4, Strategies 1–5, Strategies 3–4 and Strategies 3–5) also show some very similar performances.

Table 3 Consonance and dissonance scale [8]

Interval of $\mu_{C,C'}$	Meaning
[0.00-0.05]	Strong negative consonance
(0.05-0.15]	Negative consonance
(0.15-0.25]	Weak negative consonance
(0.25-0.33]	Weak dissonance
(0.33-0.43]	Dissonance
(0.43-0.57]	Strong dissonance
(0.57-0.67]	Dissonance
(0.67-0.75]	Weak dissonance
(0.75-0.85]	Weak positive consonance
(0.85-0.95]	Positive consonance
(0.95-1.00]	Strong positive consonance

Table 4 Index matrix for $\mu_{C,C'}$

	C_1	C_2	C_3	C_4	C_5	C_6	C_7	C_8	C_9	C_{10}	C_{11}	C_{12}
C_1	1.000	0.042	0.016	0.079	1.000	1.000	0.037	0.037	0.026	0.026	0.026	0.026
C_2	0.042	1.000	0.642	0.742	0.042	0.042	0.826	0.826	0.637	0.621	0.737	0.737
C_3	0.016	0.642	1.000	0.670	0.016	0.016	0.653	0.653	0.889	0.884	0.695	0.695
C_4	0.079	0.742	0.670	1.000	0.079	0.079	0.758	0.758	0.705	0.700	0.821	0.821
C_5	1.000	0.042	0.016	0.079	1.000	1.000	0.037	0.037	0.026	0.026	0.026	0.026
C_6	1.000	0.042	0.016	0.079	1.000	1.000	0.037	0.037	0.026	0.026	0.026	0.026
C_7	0.037	0.826	0.653	0.758	0.037	0.037	1.000	1.000	0.695	0.679	0.800	0.800
C_8	0.037	0.826	0.653	0.758	0.037	0.037	1.000	1.000	0.695	0.679	0.800	0.800
C_9	0.026	0.637	0.889	0.705	0.026	0.026	0.695	0.695	1.000	0.984	0.753	0.753
C_{10}	0.026	0.621	0.884	0.700	0.026	0.026	0.679	0.679	0.984	1.000	0.747	0.747
C_{11}	0.026	0.737	0.695	0.821	0.026	0.026	0.800	0.800	0.753	0.747	1.000	1.000
C_{12}	0.026	0.737	0.695	0.821	0.026	0.026	0.800	0.800	0.753	0.747	1.000	1.000

The criteria pairs that still in a consonance (strong positive consonance, positive consonance or weak positive consonance) are pairs between criteria C_2, C_3, C_4 and C_7, C_8, ..., C_{12}. They correspond to the Strategies 1, 2 and 3, and combinations of them. In all this strategies there are forbidden regions, therefore the ACO algorithm performs in a similar way when we apply some of these strategies.

The criteria pairs with value of $\mu_{C,C'} = [0.75 - 0.25)$ are in dissonance, i.e. there are not any dependencies between these criteria. According considered scale (Table 3) these criteria are independent. The ACO algorithm with random strategies and the ACO algorithm with strategies with forbidden regions perform in a very different way, thus we can not find any relation between them.

Table 5 Index matrix for $\nu_{C,C'}$

	C_1	C_2	C_3	C_4	C_5	C_6	C_7	C_8	C_9	C_{10}	C_{11}	C_{12}
C_1	0.000	0.000	0.000	0.000	0.000	0.000	0.000	0.000	0.000	0.000	0.000	0.000
C_2	0.000	0.000	0.300	0.137	0.000	0.000	0.095	0.095	0.295	0.311	0.195	0.195
C_3	0.000	0.300	0.000	0.237	0.000	0.000	0.295	0.295	0.079	0.079	0.263	0.263
C_4	0.000	0.137	0.237	0.000	0.000	0.000	0.137	0.137	0.189	0.195	0.084	0.084
C_5	0.000	0.000	0.000	0.000	0.000	0.000	0.000	0.000	0.000	0.000	0.000	0.000
C_6	0.000	0.000	0.000	0.000	0.000	0.000	0.000	0.000	0.000	0.000	0.000	0.000
C_7	0.000	0.095	0.295	0.137	0.000	0.000	0.000	0.000	0.242	0.258	0.137	0.137
C_8	0.000	0.095	0.295	0.137	0.000	0.000	0.000	0.000	0.242	0.258	0.137	0.137
C_9	0.000	0.079	0.079	0.189	0.000	0.000	0.242	0.242	0.000	0.016	0.205	0.205
C_{10}	0.000	0.084	0.084	0.195	0.000	0.000	0.258	0.258	0.016	0.000	0.211	0.211
C_{11}	0.000	0.263	0.263	0.084	0.000	0.000	0.137	0.137	0.205	0.211	0.000	0.000
C_{12}	0.000	0.263	0.263	0.084	0.000	0.000	0.137	0.137	0.205	0.211	0.000	0.000

Fig. 1 Presentation of ICrA results in the intuitionistic fuzzy interpretation triangle

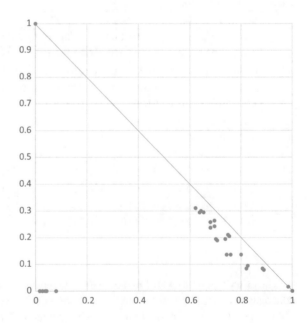

As we mentioned above the results for the criteria pairs with high $\pi_{C,C'}$-value, more than 0.9, will not be discussed. Based on these results we can not make any conclusions about the relation between different ACO performances based on considered strategies.

After ICrA application of results with different start strategies with intuitionistic fuzzy estimations we obtained the two IM with the relations between considered 12 criteria. The resulting IMs for $\mu_{C,C'}$ and $\nu_{C,C'}$ values are shown in Tables 8 and 9.

Table 6 Values of $\mu_{C,C'}$, $\nu_{C,C'}$, $\pi_{C,C'}$ of the criteria pairs – part 1

Criteria pairs	$\mu_{C,C'}$	$\nu_{C,C'}$	$\pi_{C,C'}$
$C_1 - C_5$	1.000	0.000	0.000
$C_1 - C_6$	1.000	0.000	0.000
$C_5 - C_6$	1.000	0.000	0.000
$C_7 - C_8$	1.000	0.000	0.000
$C_{11} - C_{12}$	1.000	0.000	0.000
$C_9 - C_{10}$	0.984	0.016	0.000
$C_3 - C_9$	0.889	0.079	0.032
$C_3 - C_{10}$	0.889	0.079	0.032
$C_2 - C_7$	0.826	0.095	0.079
$C_2 - C_8$	0.826	0.095	0.079
$C_4 - C_{11}$	0.821	0.084	0.095
$C_4 - C_{12}$	0.821	0.084	0.095
$C_7 - C_{11}$	0.800	0.137	0.063
$C_7 - C_{12}$	0.800	0.137	0.063
$C_8 - C_{11}$	0.800	0.137	0.063
$C_8 - C_{12}$	0.800	0.137	0.063
$C_4 - C_7$	0.758	0.137	0.105
$C_4 - C_8$	0.758	0.137	0.105
$C_9 - C_{11}$	0.753	0.205	0.042
$C_9 - C_{12}$	0.753	0.205	0.042
$C_{10} - C_{11}$	0.747	0.211	0.042
$C_{10} - C_{12}$	0.747	0.211	0.042
$C_2 - C_4$	0.742	0.137	0.121
$C_2 - C_{11}$	0.737	0.195	0.068
$C_2 - C_{12}$	0.737	0.195	0.068
$C_4 - C_9$	0.705	0.189	0.105
$C_4 - C_{10}$	0.700	0.195	0.105
$C_3 - C_{11}$	0.695	0.263	0.042
$C_3 - C_{12}$	0.695	0.263	0.042
$C_7 - C_9$	0.695	0.242	0.063
$C_8 - C_9$	0.695	0.242	0.063
$C_8 - C_{10}$	0.679	0.258	0.063
$C_7 - C_{10}$	0.679	0.258	0.063
$C_3 - C_4$	0.670	0.237	0.084
$C_3 - C_7$	0.653	0.295	0.053
$C_3 - C_8$	0.653	0.295	0.053
$C_2 - C_3$	0.642	0.300	0.058
$C_2 - C_9$	0.637	0.295	0.068
$C_2 - C_{10}$	0.621	0.311	0.068

Table 7 Values of the $\mu_{C,C'}$, $\nu_{C,C'}$, $\pi_{C,C'}$ of the criteria pairs – part 2

Criteria pairs	$\mu_{C,C'}$	$\nu_{C,C'}$	$\pi_{C,C'}$
$C_1 - C_4$	0.079	0.000	0.921
$C_4 - C_5$	0.079	0.000	0.921
$C_4 - C_6$	0.079	0.000	0.921
$C_1 - C_2$	0.042	0.000	0.958
$C_2 - C_5$	0.042	0.000	0.958
$C_2 - C_6$	0.042	0.000	0.958
$C_1 - C_7$	0.037	0.000	0.963
$C_1 - C_8$	0.037	0.000	0.963
$C_5 - C_7$	0.037	0.000	0.963
$C_5 - C_8$	0.037	0.000	0.963
$C_6 - C_7$	0.037	0.000	0.963
$C_6 - C_8$	0.037	0.000	0.963
$C_1 - C_9$	0.026	0.000	0.974
$C_1 - C_{10}$	0.026	0.000	0.974
$C_1 - C_{11}$	0.026	0.000	0.974
$C_1 - C_{12}$	0.026	0.000	0.974
$C_5 - C_9$	0.026	0.000	0.974
$C_5 - C_{10}$	0.026	0.000	0.974
$C_5 - C_{11}$	0.026	0.000	0.974
$C_5 - C_{12}$	0.026	0.000	0.974
$C_6 - C_9$	0.026	0.000	0.974
$C_6 - C_{10}$	0.026	0.000	0.974
$C_6 - C_{11}$	0.026	0.000	0.974
$C_6 - C_{12}$	0.026	0.000	0.974
$C_1 - C_3$	0.016	0.000	0.984
$C_3 - C_5$	0.016	0.000	0.984
$C_3 - C_6$	0.016	0.000	0.984

The obtained in this case ICrA results are visualized on Fig. 2 within the specific triangular geometrical interpretation of intuitionistic fuzzy sets.

The value of $\pi_{C,C'}$ is high when one of the strategies is with only probabilistic start (controlled or not) and other is with only forbidden regions. When the two strategies are with only probabilistic start or with only forbidden strategies, the value of $\mu_{C,C'}$ is close to 1. It means that the forbidden strategies performs in a very different way than the probabilistic strategies. When one of the strategies is combination of forbidden and probabilistic, the value of $\pi_{C,C'}$ is not so high, than in a previous cases, it means that there is some similarity of the algorithms performance in this case.

Table 8 Index matrix for $\mu_{C,C'}$ (intuitionistic fuzzy estimations)

	C_1	C_2	C_3	C_4	C_5	C_6	C_7	C_8	C_9	C_{10}	C_{11}	C_{12}
C_1	1.000	0.291	0.145	0.182	1.000	1.000	0.382	0.382	0.382	0.382	0.327	0.327
C_2	0.291	1.000	0.327	0.509	0.291	0.291	0.364	0.364	0.364	0.364	0.509	0.509
C_3	0.145	0.327	1.000	0.564	0.145	0.145	0.382	0.382	0.564	0.564	0.600	0.600
C_4	0.182	0.509	0.564	1.000	0.182	0.182	0.309	0.309	0.291	0.291	0.564	0.564
C_5	1.000	0.291	0.145	0.182	1.000	1.000	0.382	0.382	0.382	0.382	0.327	0.327
C_6	1.000	0.291	0.145	0.182	1.000	1.000	0.382	0.382	0.382	0.382	0.327	0.327
C_7	0.382	0.364	0.382	0.309	0.382	0.382	1.000	1.000	0.600	0.600	0.473	0.473
C_8	0.382	0.364	0.382	0.309	0.382	0.382	1.000	1.000	0.600	0.600	0.473	0.473
C_9	0.382	0.364	0.564	0.291	0.382	0.382	0.600	0.600	1.000	1.000	0.527	0.527
C_{10}	0.382	0.364	0.564	0.291	0.382	0.382	0.600	0.600	1.000	1.000	0.527	0.527
C_{11}	0.327	0.509	0.600	0.564	0.327	0.327	0.473	0.473	0.527	0.527	1.000	1.000
C_{12}	0.327	0.509	0.600	0.564	0.327	0.327	0.473	0.473	0.527	0.527	1.000	1.000

Table 9 Index matrix for $\nu_{C,C'}$

	C_1	C_2	C_3	C_4	C_5	C_6	C_7	C_8	C_9	C_{10}	C_{11}	C_{12}
C_1	0.000	0.000	0.000	0.000	0.000	0.000	0.000	0.000	0.000	0.000	0.000	0.000
C_2	0.000	0.000	0.492	0.236	0.000	0.000	0.218	0.218	0.218	0.218	0.273	0.273
C_3	0.000	0.491	0.000	0.327	0.000	0.000	0.345	0.345	0.164	0.164	0.182	0.182
C_4	0.000	0.236	0.327	0.000	0.000	0.000	0.382	0.382	0.364	0.364	0.182	0.182
C_5	0.000	0.000	0.000	0.000	0.000	0.000	0.000	0.000	0.000	0.000	0.000	0.000
C_6	0.000	0.000	0.000	0.000	0.000	0.000	0.000	0.000	0.000	0.000	0.000	0.000
C_7	0.000	0.218	0.345	0.382	0.000	0.000	0.000	0.000	0.182	0.182	0.145	0.145
C_8	0.000	0.218	0.345	0.382	0.000	0.000	0.000	0.000	0.182	0.182	0.145	0.145
C_9	0.000	0.218	0.164	0.364	0.000	0.000	0.182	0.182	0.000	0.000	0.018	0.018
C_{10}	0.000	0.218	0.164	0.364	0.000	0.000	0.182	0.182	0.000	0.000	0.018	0.018
C_{11}	0.000	0.273	0.182	0.182	0.000	0.000	0.145	0.145	0.018	0.018	0.000	0.000
C_{12}	0.000	0.273	0.182	0.182	0.000	0.000	0.145	0.145	0.018	0.018	0.000	0.000

Again for better understanding of the results the values of the $\mu_{C,C'}$, $\nu_{C,C'}$, $\pi_{C,C'}$ of the criteria pairs are sorted by the value of the $\mu_{C,C'}$. The list is presented in Tables 10 and 11. Table 10 shows the criteria pair starting with high degrees of "agreement" ($\mu_{C,C'}$) and low value for the degree of "disagreement" ($\nu_{C,C'}$).

We observe that for all pairs of criteria, the degree of disagreement is low. It means that there are not pairs of strategies where the algorithm performance to be in an opposite way. Table 11 shows the criteria pair with high degree of "uncertainty". Comparing with the results from Tables 8 and 9 we can conclude that the intuitionistic fuzzy estimation gives more realistic estimation.

Fig. 2 Presentation of ICrA results (intuitionistic fuzzy estimations) in the intuitionistic fuzzy interpretation triangle

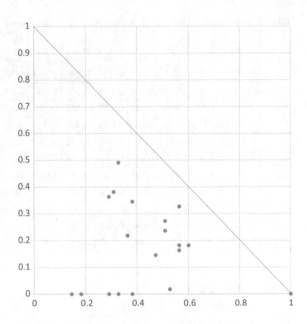

6　Conclusion

In this paper an ICrA of ACO performance with different start strategies is considered. The ACO algorithm with start strategies is tested on MKP as a representative of subset problems. The ICrA is applied to establish the relations and dependencies between the ACO performance based on different start strategies. The studied start strategies are 12. Part of them use disallowance of some of the regions of the search space for one or more iterations. The obtained ICrA results are discussed on the basis of the scale for defining consonance and dissonance between the considered criteria. We can conclude that criteria corresponding to the strategies without forbidden regions are in positive consonance, as well as the criteria corresponding to the strategies with forbidden regions. The criteria corresponding to the strategies with forbidden regions are in dissonance with criteria corresponding to strategies without forbidden regions. Comparing fuzzy estimation with intuitionistic fuzzy estimation, the intuitionistic fuzzy estimation gives more realistic results. When we apply intuitionistic fuzzy estimation the ICrA shows some even small similarity of the algorithm performance between only random strategies and strategies which are combination of probabilistic and forbidden.

Table 10 Values of the $\mu_{C,C'}$, $\nu_{C,C'}$, $\pi_{C,C'}$ of the criteria pairs – part 1 (Intuitionistic fuzzy estimations)

Criteria pairs	$\mu_{C,C'}$	$\nu_{C,C'}$	$\pi_{C,C'}$
$C_1 - C_5$	1.000	0.000	0.000
$C_1 - C_6$	1.000	0.000	0.000
$C_5 - C_6$	1.000	0.000	0.000
$C_7 - C_8$	1.000	0.000	0.000
$C_9 - C_{10}$	1.000	0.000	0.000
$C_{11} - C_{12}$	1.000	0.000	0.000
$C_3 - C_{11}$	0.600	0.182	0.218
$C_3 - C_{12}$	0.600	0.182	0.218
$C_7 - C_9$	0.600	0.182	0.218
$C_7 - C_{10}$	0.600	0.182	0.218
$C_8 - C_9$	0.600	0.182	0.218
$C_8 - C_{10}$	0.600	0.182	0.218
$C_3 - C_4$	0.564	0.327	0.109
$C_3 - C_9$	0.564	0.164	0.273
$C_3 - C_{10}$	0.564	0.164	0.273
$C_4 - C_{11}$	0.564	0.182	0.255
$C_4 - C_{12}$	0.564	0.182	0.255
$C_9 - C_{11}$	0.527	0.018	0.455
$C_9 - C_{12}$	0.527	0.018	0.455
$C_{10} - C_{11}$	0.527	0.018	0.455
$C_{10} - C_{12}$	0.527	0.018	0.455
$C_2 - C_4$	0.509	0.236	0.255
$C_2 - C_{11}$	0.509	0.273	0.218
$C_2 - C_{12}$	0.509	0.273	0.218
$C_7 - C_{11}$	0.473	0.145	0.382
$C_7 - C_{12}$	0.473	0.145	0.382
$C_8 - C_{11}$	0.473	0.145	0.382
$C_8 - C_{12}$	0.473	0.145	0.382
$C_1 - C_7$	0.382	0.000	0.618
$C_1 - C_8$	0.382	0.000	0.618
$C_1 - C_9$	0.382	0.000	0.618
$C_1 - C_{10}$	0.382	0.000	0.618
$C_3 - C_7$	0.382	0.345	0.273

Table 11 Values of the $\mu_{C,C'}$, $\nu_{C,C'}$, $\pi_{C,C'}$ of the criteria pairs – part 2 (Intuitionistic fuzzy estimations)

Criteria pairs	$\mu_{C,C'}$	$\nu_{C,C'}$	$\pi_{C,C'}$
$C_3 - C_8$	0.382	0.345	0.273
$C_5 - C_7$	0.382	0.000	0.618
$C_5 - C_8$	0.382	0.000	0.618
$C_5 - C_9$	0.382	0.000	0.618
$C_5 - C_{10}$	0.382	0.000	0.618
$C_6 - C_7$	0.382	0.000	0.618
$C_6 - C_8$	0.382	0.000	0.618
$C_6 - C_9$	0.382	0.000	0.618
$C_6 - C_{10}$	0.382	0.000	0.618
$C_2 - C_7$	0.364	0.218	0.418
$C_2 - C_8$	0.364	0.218	0.418
$C_2 - C_9$	0.364	0.218	0.418
$C_2 - C_{10}$	0.364	0.218	0.418
$C_1 - C_{11}$	0.327	0.000	0.673
$C_1 - C_{12}$	0.327	0.000	0.673
$C_2 - C_3$	0.327	0.491	0.182
$C_5 - C_{11}$	0.327	0.000	0.673
$C_5 - C_{12}$	0.327	0.000	0.673
$C_6 - C_{11}$	0.327	0.000	0.673
$C_6 - C_{12}$	0.327	0.000	0.673
$C_4 - C_7$	0.309	0.382	0.309
$C_4 - C_8$	0.309	0.382	0.309
$C_1 - C_2$	0.291	0.000	0.709
$C_2 - C_5$	0.291	0.000	0.709
$C_2 - C_6$	0.291	0.000	0.709
$C_4 - C_9$	0.291	0.364	0.345
$C_4 - C_{10}$	0.291	0.364	0.345
$C_1 - C_4$	0.182	0.000	0.818
$C_4 - C_5$	0.182	0.000	0.818
$C_4 - C_6$	0.182	0.000	0.818
$C_1 - C_3$	0.145	0.000	0.855
$C_3 - C_5$	0.145	0.000	0.855
$C_3 - C_6$	0.145	0.000	0.855

Acknowledgements Work presented here is partially supported by the National Scientific Fund of Bulgaria under Grants DFNI-02-5/2014 "Intercriteria Analysis – A Novel Approach to Decision Making" and DFNI I02/20 "Efficient Parallel Algorithms for Large Scale Computational Problems", and by the Polish-Bulgarian collaborative Grant "Parallel and Distributed Computing Practices".

References

1. Angelova, M., Roeva, O., Pencheva, T.: InterCriteria analysis of crossover and mutation rates relations in simple genetic algorithm. Proc. Fed. Conf. Comput. Sci. Inf. Syst. **5**, 419–424 (2015)
2. Atanassov, K.: On index matrices, part 1: standard cases. Adv. Stud. Contemp. Math. **20**(2), 291–302 (2010)
3. Atanassov, K.: On Intuitionistic Fuzzy Sets Theory. Springer, Berlin (2012)
4. Atanassov, K., Mavrov, D., Atanassova, V.: Intercriteria decision making: a new approach for multicriteria decision making. Based Index Matrices Intuitionistic Fuzzy Sets. Issues IFSs GNs **11**, 1–8 (2014)
5. Atanassova, V., Mavrov, D., Doukovska, L., Atanassov, K.: Discussion on the threshold values in the intercriteria decision making approach. Notes on Intuitionistic Fuzzy Sets **20**(2), 94–99 (2014)
6. Atanassova, V., Doukovska, L., Atanassov, K., Mavrov, D.: Intercriteria decision making approach to EU member states competitiveness analysis. In: Proceedings of the International Symposium on Business Modeling and Software Design - BMSD'14, pp. 289–294 (2014)
7. Atanassova, V., Doukovska, L., Karastoyanov, D., Capkovic, F.: Intercriteria decision making approach to EU member states competitiveness analysis: trend analysis, In: Angelov, P., et al. (eds.) Intelligent Systems' 2014. Advances in Intelligent Systems and Computing, vol. 322, pp. 107–115 (2014)
8. Atanassov, K., Atanassova, V., Gluhchev, G.: InterCriteria analysis: ideas and problems. Notes Intuitionistic Fuzzy Sets **21**(1), 81–88 (2015)
9. Diffe, W., Hellman, M.E.: New direction in cryptography. IEEE Trans. Inf. Theory **IT-36**, 644–654 (1976)
10. Dorigo, M., Stutzler, T.: Ant Colony Optimization. MIT Press, Cambridge (2004)
11. Dorigo, M., Birattari, M.: Ant colony optimization. In: Sammut, C., Webb, G.I (eds.) Encyclopedia of Machine Learning, pp. 36–39. Springer, Berlin (2010)
12. Fidanova, S.: Evolutionary algorithm for multiple knapsack problem. In: Proceedings of International Conference Parallel Problems Solving from Nature, Real World Optimization Using Evolutionary computing, Granada, Spain (2002)
13. Fidanova, S., Atanassov, K., Marinov, P.: Generalized Nets and Ant Colony Optimization. Academic Publishing House, Bulgarian Academy of Sciences (2011)
14. Gendreau, M., Potvin, J.-Y.: Handbook of Metaheuristics. International Series in Operations Research and Management Science. Springer, Berlin (2010)
15. Kochemberger, G., McCarl, G., Wymann, F.: Heuristic for general inter programming. J. Decision Sci. **5**, 34–44 (1974)
16. Lessing, L., Dumitrescu, I., Stutzle, T.: A Comparison Between ACO Algorithms for the Set Covering Problem, Ant Colony Optimization and Swarm Intelligence. Lecture Notes in Computer Science, vol. 3172. Springer, Germany (2004)
17. Martello, S., Toth, P.: A mixtures of dynamic programming and branch-and-bound for the subset-sum problem. Manag. Sci. **30**, 756–771 (1984)
18. Reiman, M., Loumanns, M.: A hybride ACO algorithm for a capacitated minimum spanning tree problem. In: Proceedings of First International Workshop on Hybrid Metaheuristics, Valencia, Spain, pp. 1–10 (2004)

19. Roeva, O., Fidanova, S., Vassilev, P., Gepner, P.: Intercriteria analysis of a model parameters identification using genetic algorithm. Proc. Fed. Conf. Comput. Sci. Inf. Syst. **5**, 501–506 (2015)
20. Roeva, O., Fidanova, S., Paprzycki, M.: Intercriteria analysis of ACO and GA hybrid algorithms. Stud. Comput. Intell. **610**, 107–126 (2016)
21. Sinha, A., Zoltner, A.A.: The multiple-choice knapsack problem. J. Op. Res. **27**, 503–515 (1979)
22. Stutzle, T., Dorigo, M.: ACO algorithm for the traveling salesman problem. Evolutionary Algorithms in Engineerings and Computer Science, pp. 163–183. Wiley, New York (1999)
23. Todinova, S., Mavrov, D., Krumova, S., Marinov, P., Atanassova, V., Atanassov, K., Taneva, S.G.: Blood plasma thermograms dataset analysis by means of intercriteria and correlation analyses for the case of colorectal cancer. Int. J. Bioautomation **20**(1), 115–124 (2016)
24. Zhang, T., Wang, S., Tian, W., Zhang, Y.: ACO-VRPTWRT: a new algorithm for the vehicle routing problems with time windows and re-used vehicles based on ant colony optimization. In: Proceedings of Sixth International Conference on Intelligent Systems Design and Applications, pp. 390–395. IEEE Press (2006)

Evolutionary Approach for Tuning
of Longwall Scraper Conveyor Model

Piotr Przystałka and Andrzej Katunin

Abstract The modeling of machines and their operation modes is a key approach for optimization of their performance as well as for avoiding unwanted operational states which may lead to the occurrence of faults, and finally, to the breakdown. The developed model of a machine should be always parametrized, i.e. the certain number of parameters should be selected in the certain ranges. The most of the parameters can be selected based on engineering documentation and experts' knowledge, however, for some of them this knowledge cannot be directly acquired which leads to the parameter uncertainty. One of the approaches allowing selection of these uncertain parameters is a tuning procedure of a model. The paper deals with a heuristic optimization method for automatic tuning of key parameters of longwall scraper conveyor model. In the first part of the paper, the analytical model and simulator of the conveyor system are described as well as the evolutionary algorithm for tuning this model is proposed. In the case study, the merits and limitations of the evolutionary approach are analysed. The obtained results prove that the proposed tuning method has high practical potential and it may be applied in real mining conditions.

1 Introduction

The longwall scraper conveyors are the machines used in the mechanized underground coal mines for transporting a coal. Since these machines usually work in extremely difficult operational conditions it is essential to monitor their performance in order to prevent unwanted operational modes and machinery downtime as well as to implement the knowledge obtained from the monitoring process into the re-designing processes which allows increasing their reliability, effectiveness and safety. However, considering the operational conditions in the underground coal mines as well as difficulties in physical access to sources of various signals and difficulties with their measurement, the monitoring of physical working parameters is often limited

P. Przystałka · A. Katunin (✉)
Silesian University of Technology, Institute of Machinery Design,
18A Konarskiego Str., 44-100 Gliwice, Poland
e-mail: andrzej.katunin@polsl.pl

© Springer International Publishing AG 2018
S. Fidanova (ed.), *Recent Advances in Computational Optimization*,
Studies in Computational Intelligence 717, DOI 10.1007/978-3-319-59861-1_5

to few main quantities, which causes that the measurement data is incomplete, and makes the diagnosis and prognosis of these machines difficult. In order to predict an inappropriate behavior of a conveyor and prevent its unwanted operational modes, it is essential to develop a simulator (or mathematical/numerical model) which allows testing various operational scenarios, including the occurrence of various types of faults.

The model-based approach in diagnostics and condition monitoring of underground mines machinery is widely applicable in numerous industrial solutions. To date, many of such models were developed for scraper and belt conveyors. From the variety of types of such models one can distinguish analytical models of dynamical behavior of such conveyors, and models based on numerical formulations like Finite Element Method or Discrete Element Method (see e.g. [1]). Numerous analytical models which describe motion and dynamics of longwall scraper conveyors were studied by the team of Dolipski [2–4]. Similar studies on analytical modeling of mining conveyors were performed by Mao et al. [5], where the authors developed a dynamic mathematical model of a scraper conveyor in order to investigate the chain behavior under certain operation conditions; Zhang and Meng [6], where the scraper conveyor sprocket transmission system was simulated in order to analyze dynamic loading of a conveyor during transporting of excavated material; Eshin [7], where the dynamic behavior of a scraper conveyor was simulated in Matlab®/Simulink® environment in order to test a performance of dynamic control algorithms; Fan et al. [8], where the authors developed procedures of coordination control of a shearer, dynamic support and a scraper conveyor; and finally Cenacewicz and Przystałka [9], Cenacewicz and Katunin [10] developed the models of belt and scraper conveyors for evaluation of their dynamical behavior under certain operational scenarios. An interesting study was performed by Herbuś et al. [11], where the authors developed a model in order to perform numerical simulations for developing generalized and parametrized control algorithm for a scraper conveyor.

This short survey shows wideness and importance of development of mathematical models for diagnosis and control purposes of mining machinery. However, in the most cases, modeling of dynamic behavior of such machinery is difficult, since many operational parameters are not available or even not measurable or impossible to acquire. This leads to the incompleteness and uncertainty of a developed model. Thus, in order to achieve such parameters one can perform simplifications of the model, assume them basing on literature data and experts' knowledge, or use knowledge discovery and optimization techniques. Currently, the most common approach in this task is the manual adjustment of values of behavioural parameters of the model taking into account the data included in technical documentation or in domain literature as well as domain expert's knowledge. Obviously, such an approach is ineffective and leads to the increasing error with an increase of number of uncertain parameters and overall level of model uncertainty. On the other hand, in the recent years, heuristic methods based on the natural phenomena of evolution, such as simulated annealing algorithm, genetic algorithms, differential evolution, harmony or tabu search ideas, swarm-inspired methods, etc. have been developed and applied to model and solve real-life global optimization problems [12]. Furthermore, this kind of optimization algorithms has long been applied for tuning values of unknown parameters of different types of models [13, 14].

As one can observe, the problem of unknown parameter estimation is not enough discussed in many studies related to analytical modeling of mining conveyors, see e.g. [3, 4, 10]. To the best knowledge of the authors of this paper, this is one of the first attempts on applying a heuristic optimization technique for automatic tuning of parameters of longwall scraper conveyor model. The main goal of this study is the introduction of the new approach based evolutionary tuning of the mathematical model of the longwall scraper conveyor. This approach allows for adjusting the values of uncertain parameters of the model, and thus make it fully defined. This, in turn, allows for using this model for modeling of various scenarios of operation as well as for diagnosing the considered machine using the model-based diagnostics approach. The tuned model was validated, and the validation studies confirmed the effectiveness of the proposed evolutionary approach.

The paper is composed as follows. In Sect. 2 we present an analytical model and simulator of longwall scraper conveyor. Section 3 deals with the evolutionary approach for tuning of conveyor model. Section 4 shows the pros and cons of the proposed tuning method. The paper ends with conclusions on development of the model, tuning procedure as well as further directions of research.

2 Analytical Model and Simulator of Longwall Scraper Conveyor

The mathematical model of a scraper conveyor consists of the following submodels: model of a doubled main drive, model of auxiliary drive, model of mine breaker and contactor control, power supply model, model of equations of motion, model of masses, and model of motion resistance. A detailed mathematical and phenomenological description can be found in [10]. In this study we limit to the expressions that reflect the tuned parameters only.

The scraper conveyor was modeled as a 4-segment discretised model similar to the approach used by Dolipski [2] (see Fig. 1). The equations of motion for this model are given by:

$$\dot{V}(j) m(j) l(j) = F(j+1) - F(j) + P(j) - W(j) l(j), \qquad (1)$$

$$\dot{F}(j+1) l(j) = AE[V(j+1) - V(j)], \qquad (2)$$

where j denotes j-th segment of a conveyor, $V(j)$ and $F(j)$ are the velocity and force in j-th segment of a conveyor, respectively (dots on top denote derivatives), $l(j)$ and $m(j)$ are the length and the mass of j-th segment, respectively, $W(j)$ denotes motion resistance of j-th segment, $P(j)$ is the internal force in j-th segment, and A and E are the area of a cross-section and modulus of elasticity of a chain, respectively.

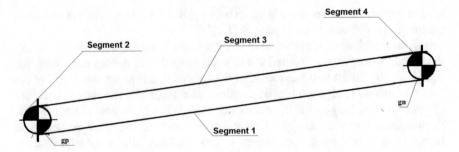

Fig. 1 The discretised model of a longwall scraper conveyor

In Eqs. (1) and (2) $m(j)$ represents a submodel of masses for tendons and sprockets. Following [2] the model of masses for operating tendon is described by:

$$m_{xx}(j) = \left(m_k + \frac{m_z}{2p_z} + c_{u,xx}(j)\, m_{u,xx}(j) \right) L, \tag{3}$$

where $m_{xx}(j)$ – the mass of j-th segment of a conveyor, m_k – the unitary mass of a chain, m_z – the mass of a scraper, p_z – pitch of bearing elements, $c_{u,xx}(j)$ – mass coefficient for chain vibration, $m_{u,xx}(j)$ – unitary mass of excavated material, and L is the whole length of a conveyor.

For non-operating tendon (3) simplifies to the following form:

$$m_{xx}(j) = \left(m_k + \frac{m_z}{2p_z} \right) L. \tag{4}$$

In the case of sprockets the masses can be determined using the following equations:

$$m_{gnnap}(j) = \left(m_k + \frac{m_z}{2p_z} + \frac{J_{nnap}(j)}{l(j)\, r_{nnap}^2(j)} \right) L, \tag{5}$$

$$m_{gnap}(j) = \left(m_k + \frac{m_z}{2p_z} + \frac{1}{2l(j)\, r_b^2(j)} \left(J_b(j) \right. \right.$$
$$\left. \left. + i_p^2(j)\, \eta_p(j) + i_p^2(j+1)\, \eta_p(j+1)\, J_p(j+1) \right) \right) L, \tag{6}$$

where $m_{gnnap}(j)$, $m_{gnap}(j)$ – equivalent masses of the j-th segment of a turning and driving sprockets, respectively, r_b – radius of a sprocket, i_p – gear ratio, J_b – moment of inertia of a sprocket, η_p – gear efficiency.

The motion resistance, based on [2], can be described as follows:

$$W_{xx}(j) = g m_{xx}(j)\, a_{xx}(j), \tag{7}$$

where g is the gravitational acceleration, and $a_{xx}(j)$ is the coefficient of approximation of external friction.

The torque losses which should be considered in the model can be determined based on the following relationship:

$$\Delta M_p = \frac{M_b (1 - \eta_p)}{\omega_0 \eta_p}, \tag{8}$$

where ΔM_p denotes the torque losses of a gear, M_b is the torque of the drive drum, and η_p is the gear efficiency. The engine torque relegated to the drive shaft of the driving sprocket can be determined basing on the following relationship [15]:

$$M_n = (M_s - \Delta M_p) \eta i, \tag{9}$$

where M_s is the drive torque of the engine relegated to the drive shaft sprocket, and $\eta = \eta_s \eta_p$, η_s is the efficiency of the coupling.

The output torque can be described as:

$$M_{\text{out}} = (M_e + J_s) \frac{d\omega_s}{dt}, \tag{10}$$

where J_s is the moment of inertia of the motor, ω_s is the angular velocity of the motor, and M_e is the elastic torque given by:

$$M_e = c\phi + \mu \frac{d\phi}{dt}, \tag{11}$$

where c is the elasticity coefficient, ϕ is the angle of rotation, and μ is the damping coefficient.

Basing on the described mathematical model the of a scraper conveyor was developed and implemented in the Matlab®/Simulink® environment. The model assumes the ability to simulate different operational scenarios of a conveyor, and the ability to simulate operational faults typical for scraper conveyors working underground. The simulator consists of several subsystems, related to the above-described mathematical model, which are responsible for various actions and operational scenarios.

3 Evolutionary Algorithm for Tuning of Conveyor Model

The dynamic behavior of the described model strongly depends on values of key parameters corresponding to physical properties of the real conveyor system. The total number of these parameters can be declared as D. As it is mentioned above, the most of them can be easily established in the direct way because there is the possibility for gauging and quantifying physical properties of the plant. On the other hand, the rest values declared as $\mathbf{x} = [x_1 \ x_2 \ \ldots \ x_d]$ can be indirectly found using signals of process variables which are collected during monitoring of the object and simulation of the model. We assume that the number of observed signals (real and simulated) is equal to J, whereas the number of experiments (scenarios) is I. Henceforth, real (r) and modeled (m) time series that are needed for tuning purposes may be denoted as

$$\mathbf{y}_r^{ij} = \left[y_r^{ij}(1) \ y_r^{ij}(2) \ \ldots \ y_r^{ij}(K) \right],$$

and

$$\mathbf{y}_m^{ij}(\mathbf{x}) = \left[y_m^{ij}(1, \mathbf{x}) \ y_m^{ij}(2, \mathbf{x}) \ \ldots \ y_m^{ij}(K, \mathbf{x}) \right],$$

where j is the j-th signal (real or artificial) collected in the i-th experiment scenario, K is the number of samples.

The main objective of the tuning procedure is to adjust the values of the parameters $x_1, x_2, \ldots, x_d$ in order to obtain the smallest difference between the response of the system and the response predicted by the proposed model for each scenario. Therefore, the optimization problem can be written as follows:

$$\text{Minimize } C(\mathbf{x}) = f\left[\mathbf{y}_r^{ij}, \mathbf{y}_m^{ij}(\mathbf{x}), I, J, K \right]$$
$$\text{subject to } \Omega(\mathbf{x}), \tag{12}$$

where f represents a function (with constant arguments I, J, K) for comparison of two time series, whilst Ω denotes boundaries and constraints in the optimization process. The optimal solution $\mathbf{x}^*$ is found if the criterion function C has a relative minimum value at $\mathbf{x} = \mathbf{x}^*$, that means if

$$\mathbf{x}^* = \arg \min_{\mathbf{x} \in \Omega} C(\mathbf{x}). \tag{13}$$

As to be expected, the criterion function C can be formulated in several ways. In this study, the authors propose two variants of this function. The first one is prepared applying the Minkowski distance of order p

$$C(\mathbf{x}, p) = \sum_{i=1}^{I} \sum_{j=1}^{J} \sqrt[p]{\sum_{k=1}^{K} \left| y_r^{ij}(k) - y_m^{ij}(k, \mathbf{x}) \right|^p}. \tag{14}$$

This measure is a metric in a normed vector space which is considered as a generalization of both the Euclidean distance and the Manhattan distance. The second function is composed of three sub-criteria

$$C(\mathbf{x}, \mathbf{w}) = w_1 C_1(\mathbf{x}) + w_2 C_2(\mathbf{x}) + w_3 C_3(\mathbf{x}), \tag{15}$$

where w_1, w_2 and w_3 are used in order to control the significance of each component. The first criterion function in Eq. (15) is grounded on the mean absolute percent error and therefore it can be written as

$$C_1(\mathbf{x}) = \frac{100}{IJK} \sum_{i=1}^{I} \sum_{j=1}^{J} \sum_{k=1}^{K} \left| \frac{y_r^{ij}(k) - y_m^{ij}(k, \mathbf{x})}{z^j} \right|, \tag{16}$$

where z^j corresponds to the range of the j-th sensor.

The second function is declared making use of cross-correlation as the base of a measure of the total lag τ between real and simulated signals

$$C_2(\mathbf{x}) = \tau = \sum_{i=1}^{I} \sum_{j=1}^{J} \tau^{ij} \left[R_{xy} \left(y_r^{ij}, y_m^{ij}(\mathbf{x}) \right) \right]. \tag{17}$$

The last component is used in order to find the aggregate value of the maximum absolute errors which are present in signals collected under all scenarios

$$C_3(\mathbf{x}) = \sum_{i=1}^{I} \sum_{j=1}^{J} \max \left| \frac{\mathbf{y}_r^{ij} - \mathbf{y}_m^{ij}(\mathbf{x})}{z^j} \right|. \tag{18}$$

The solution of the optimization problem can be found using a limited number of strategies. Derivative-based approaches cannot be employed in this paper, mainly due to the form of the objective function C. Moreover, one can easily observe, that the return value of this function depends on the measurement noise in the real-world data as well as the virtual measurements (with simulated disturbances, noise and computing errors as well) obtained during numerical computations. In contrast, pure stochastic optimization methods, for example, Monte Carlo techniques will not be able to find an accurate solution with guaranteeing polynomial-time convergence because of the time of numerical computations of the conveyor model. Therefore, the authors decided to use the evolutionary algorithm which is known as one of the most common heuristic optimization methods. The general scheme of the evolutionary optimization strategy for tuning of parameters of longwall scraper conveyor model can be viewed in Fig. 2.

Evolutionary algorithms are based on the natural selection process that mimics biological evolution. In order to apply such an optimization technique for finding a solution of the problem it is necessary to define the following properties of the algorithm [16]: the representation of the individuals, the fitness function, selection and succession methods, crossover and mutation operators. It is assumed, that the number of individuals in the population is fixed at each epoch of the evolutionary process and that individuals are composed of genes representing real numeric values of adjustable model parameters. The length of the chromosome is dependent on the number of these parameters and equals the length of the vector $\mathbf{x}$:

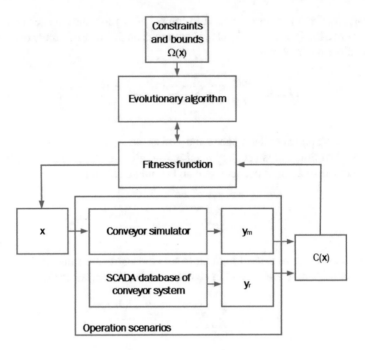

Fig. 2 General scheme of the optimization strategy used for tuning purposes

$$\mathbf{chr} = \mathbf{x} = [x_1 \ x_2 \ \ldots \ x_d]. \tag{19}$$

The initial population is generated using the normal distribution. The fitness value of an individual is computed using the objective function (14 or 15). The best fitness value for a population is the smallest fitness value for every individual in the population. Stochastic uniform is applied to choose parents for the next generation, whereas succession operations are realized by defining the reproduction rules characterized by two parameters: elite count (δ_s) and crossover fraction (p_c). The first parameter is the number of individuals with the best fitness values in the current generation that are guaranteed to survive to the next generation. The second one is the fraction of individuals in the next generation, other than elite children, that are created by crossover. It is decided to use a simple heuristic crossover operator. On the basis of two individuals $\mathbf{x}_1$ and $\mathbf{x}_2$, in the case if $C(\mathbf{x}_1) < C(\mathbf{x}_2)$, then the new one $\mathbf{x}_3$ is created according to the formula mentioned below

$$\mathbf{x}_3 = \mathbf{x}_2 + \lambda_h (\mathbf{x}_1 - \mathbf{x}_2), \tag{20}$$

where λ_h is a fraction pointing at the better adapted individual.

A mutation operator is responsible for generating heterogeneous individuals. This operator is based on an adaptive feasible method and it randomly generates directions with respect to the last successful or unsuccessful generation. The mutation operator finds a direction and step length satisfying bounds and linear constraints.

The algorithm is also described by two important parameters such as the population size and the number of generations. The values of theses parameters are arbitrarily selected during optimization experiments.

4 Case Study

4.1 Description of JOY BLS Conveyor and Its Operating Scenarios

The developed mathematical model was based on construction and parameters of the scraper conveyor of type JOY® BLS with the doubled and auxiliary drives. The parametrization of the described simulator was performed based on technical documentation of the modeled conveyor, and data available in the literature [2, 17]. The developed model is characterised by nearly fifty adjustable parameters. The main parameters of the modeled conveyor are presented in Table 1. The operational parameters were determined theoretically or selected basing on the experts' knowledge.

The simulator of a scraper conveyor provides a possibility of simulation of eight operating scenarios, which represent the characteristic considering the operations performed during work. They consist of:

Table 1 Parameters of the conveyor JOY® BLS

Parameter	Symbol	Value	Unit
Productivity	Q_p	2160	t/h
Chain type – diameter	d_c	ϕ34	mm
Chain type – pitch	p_c	126	mm
Chain type – spacing	s_c	200	mm
Diameter of the driving sprocket	D_{gn}	200	mm
Diameter of the turning sprocket	D_{gp}	200	mm
Number of chains	n_c	2	–
Velocity of the conveyor	v_p	1.3	m/s
Length of the conveyor	l_p	250	m
Gear ratio	i	13	–
Drive power	P_n	3×250	kW
Supply voltage	U_n	500	V
Sectional area of the chain	A	$9.0792 \cdot 10^{-6}$	m^2
Young modulus of the chain	E	$2.1 \cdot 10^{11}$	Pa

- S1 – idle run-up;
- S2 – idle run-up and run-down;
- S3 – idle run-up, loading, run-down;
- S4 – idle run-up, loading, unloading, run-down;
- S5 – run-up with excavated material;
- S6 – run-up with excavated material, unloading, run-down;
- S7 – run-up with loading of excavated material;
- S8 – run-up with loading of excavated material, unloading, run-down.

The loading of excavated material is modeled as a linear increase of the coefficient k_z (when it equals 1 – the full rated loading is obtained). For the performed study four of them were selected, namely S3, S4, S5 and S8 due to the significant differences between these scenarios.

Besides the parameters presented in Table 1, some of them cannot be measured and were assumed according to literature data and technical data sheets. Therefore, it is essential to tune up the model in order to simulate its behavior during realization of considered scenarios properly. The five parameters ($d = 5$) that are subject to the tuning process in this study are as follows:

- the efficiency of the drive system: $x_1 = \eta$ [-] (see (9));
- the damping factor: $x_2 = \mu$ [-] (see (11));
- the torque losses of flexible and hydrodynamic couplings: $x_3 = \Delta M_p$ [Nm] (see (8));
- the approximation friction coefficient: $x_4 = a$ [-] (see (7));
- the unitary mass of excavated material: $x_5 = m_u$ [kg/m] (see (3)).

These parameters are selected since it is not possible to obtain their values in the direct way.

4.2   Tuning Experiments and Results

The verification tests were carried out with the assumption corresponding to the process variables collected during operational states of the machine. It was decided that only nine process variables ($J = 9$) could be used for tuning:

- the load of the engines $y_1 = M_{gn}$ [Nm];
- the torque of the engines $y_2 = M_n$ [Nm] (see (9));
- the linear velocity of the chain $y_3 = v_n$ [m/s];
- the phase currents of the first and second engine $y_4 = INZ1A$, $y_5 = INZ1B$, ..., $y_9 = INZ2C$ [A] (see [10] for details).

It was also assumed that the sampling rate of the sensors was equal to 500Hz, whereas the analog-to-digital converter resolution was set to 32bits. The noise powers of selected signals were as follows: $\sigma_{v_n} = 10\mathrm{E}{-}08$, $\sigma_{Mgn} = \sigma_{Mpg} = 100$, $\sigma_I = 10\mathrm{E}{-}02$. The reference signals $\mathbf{y}_r^{ij}$ were gathered while simulation of the model for S3, S4, S5 and S8 scenarios ($I = 4$). The optimal values of parameters were chosen as follows: $x_1^r = 0.957$, $x_2^r = 0.1$, $x_3^r = 5$, $x_4^r = 3$ and $x_5^r = 461.54$. The time of the simulation was set to 40s. The error measure in the form of

$$\delta x = \frac{100\%}{d} \sum_{i=1}^{d} \delta x_i = \frac{100\%}{d} \sum_{i=1}^{d} \left| \frac{x_i^r - x_i^*}{x_i^r} \right|, \tag{21}$$

was defined in order to evaluate the performance results.

The evolutionary algorithm implemented in Global Optimization Toolbox of Matlab® software was applied in this paper. For each optimization experiment, the boundary values of parameters were chosen taking into account literature data: $0.8 \le x_1 \le 0.99, 0 \le x_2 \le 10, 2 \le x_3 \le 8, 2 \le x_4 \le 4, 300 \le x_5 \le 570$. In the first step, the performance of the evolutionary algorithm was examined through analysing the influence of the variant of the criterion function C and the values of its parameters. The feasible population method was adapted to create a random well-dispersed initial population satisfying all constraints and bounds. Fitness scaling was done using the rank method, whereas the selection of the parents to the next generation was obtained by means of the stochastic uniform method. The elite count $\delta_s = 2$ and crossover fraction $p_c = 0.8$ were chosen. The heuristic crossover function was employed with the user-defined parameter $\lambda_h = 1.2$. The remaining individuals were mutated with the use of the adaptive feasible method. The population size was equal to 50, whilst the total number of generations was set to 20. Nine trials were performed in this part of the study:

- trials from 1 to 5: fitness function was declared using Eq. (14) for $p = \{1, 2, \ldots, 5\}$;
- trials from 6 to 9: fitness function was declared using Eq. (15) for $\mathbf{w} = [1\ 0\ 0]$, $\mathbf{w} = [0\ 1\ 0]$, $\mathbf{w} = [0\ 0\ 1]$, $\mathbf{w} = [0.9\ 0.01\ 0.09]$.

The results of experiments from this stage of the study are included in the first part of Table 2. It can be stated that in the average sense, the minor errors can be achieved using the criterion function C in the form of Eq. 14. For this function, in each case beyond the 5th trial the mean error δx was close to 10%. Nevertheless, the smallest error was reached for the second criterion function that has been declared using Eq. 15 with $\mathbf{w} = [0.9\ 0.01\ 0.09]$. Hence, this variant of the fitness function was used in the next two steps. Convergence plots of two evolutionary tuning processes are presented in Fig. 3. The first one (Fig. 3a) shows the lowest efficiency of the algorithm obtained in the trial No. 6. On the other hand, the second plot (Fig. 3b) illustrates the highest efficiency of the algorithm in the trial No. 9. Using this two figures, it is possible to compare the convergences for both trials. One can see that, in the second case the dispersion of mean scores is higher than in the first one. Moreover, Fig. 4 shows examples of simulation results obtained for different values of parameters determined by the evolutionary optimization algorithm in the 6th and 9th trial.

In the second step, the authors analysed the influence of the population size and the crossover probability on the performance of the evolutionary algorithm. In order to examine this issue six experiments were conducted:

Table 2 The errors and final values of parameters determined by the evolutionary optimization algorithm

Trial No.	x_1^* [−] δx_1 [%]	x_2^* [−] δx_2 [%]	x_3^* [Nm] δx_3 [%]	x_4^* [−] δx_4 [%]	x_5^* [kg] δx_5 [%]	δx [%]
1st step						
1	0.949 0.75	0.13 30.64	6.33 26.76	3.01 0.46	457.95 0.78	11.88
2	0.950 0.72	0.09 1.76	2.91 41.79	3.02 0.76	455.64 1.28	9.26
3	0.958 0.14	0.11 13.97	7.00 40.09	3.00 0.08	462.21 0.15	10.89
4	0.956 0.09	0.08 12.26	3.37 32.49	3.00 0.02	461.23 0.07	8.98
5	0.941 1.60	0.20 102.19	4.59 8.08	2.93 2.26	471.84 2.23	23.27
6	0.919 3.96	0.27 170.47	2.25 54.87	2.67 10.79	507.01 9.85	49.99
7	0.873 8.76	0.29 192.92	5.32 6.48	2.33 22.02	405.70 12.10	48.46
8	0.899 6.03	0.061 38.74	2.01 59.62	3.06 2.08	409.92 11.18	23.53
9	0.955 0.21	0.10 4.78	6.18 23.72	2.87 4.13	486.54 5.42	7.65
2nd step						
10	0.958 0.18	0.09 3.27	7.88 57.67	2.99 0.19	463.24 0.37	12.33
11	0.934 2.36	0.06 32.17	2.84 43.11	2.98 0.66	450.75 2.34	16.13
12	0.959 0.23	0.26 165.68	3.90 21.93	2.86 4.53	492.54 6.72	39.82
13	0.957 0.06	0.10 6.74	5.66 13.34	2.99 0.19	463.24 0.37	4.14
14	0.941 1.64	0.14 46.92	6.34 26.90	3.04 1.65	440.88 4.48	16.32
15	0.958 0.19	0.10 4.83	6.92 38.49	3.00 0.10	461.55 0.00	8.72
3rd step						
16	0.965 0.92	0.18 80.97	7.86 57.29	2.96 1.22	474.40 2.79	28.64
17	0.954 0.22	0.09 1.63	3.52 29.60	3.00 0.09	459.83 0.37	6.38
18	0.943 1.38	0.08 13.24	2.53 49.31	2.97 0.69	456.95 0.99	13.12

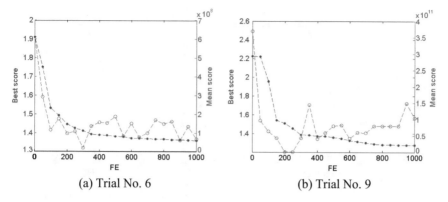

(a) Trial No. 6   (b) Trial No. 9

Fig. 3 Convergence plots of the evolutionary tuning process

- trials 10 and 11: fitness function was declared using Eq. (15) for $\mathbf{w}$ = [0.9 0.01 0.09] and the population size was equal to {30, 40}, the rest properties were the same as in the first step;
- trials from 12 to 14: fitness function was declared using Eq. (15) for $\mathbf{w}$ = [0.9 0.01 0.09] and the crossover fraction was equal to {0.6, 0.7, 0.9}, the rest properties were the same as in the first step;
- trial 15: fitness function was declared using Eq. (15) for $\mathbf{w}$ = [0.9 0.01 0.09], the population size was equal to 50 and the crossover fraction was equal to 0.7, the rest properties were the same as in the first step.

This part of the research led us to state that, the smallest error could be achieved with the use of 20 individuals in the population, whereas the crossover fraction should be set to 0.7. It is easy to observe that, these settings can provide the mean error result smaller than 5%.

In the last part of the study, the analysis of the accuracy of the evolutionary optimization was carried out in the context of the sampling rate and resolution as well as the number of the available sensors. The last three trials were done as follows:

- trial 16: fitness function was declared using Eq. (15) for $\mathbf{w}$ = [0.9 0.01 0.09], only three sensors were used v_n, $INZ1A$ and $INZ2A$, the rest properties were the same as in the second step;
- trial 17: fitness function declared using Eq. (15) for $\mathbf{w}$ = [0.9 0.01 0.09], all sensors were used, the sampling rate was equal to 1 Hz, the resolution was set to 16bits, the rest properties were the same as in the second step;
- trial 18: fitness function declared using Eq. (15) for $\mathbf{w}$ = [0.9 0.01 0.09], only three sensors were used v_n, $INZ1A$ and $INZ2A$, the sampling rate was equal to 1 Hz, the resolution was set to 16bits, the rest properties were the same as in the second step.

The last analysis is very important from a technical point of view. It can be pointed out, that in mining engineering practice it is possible to use measuring devices with

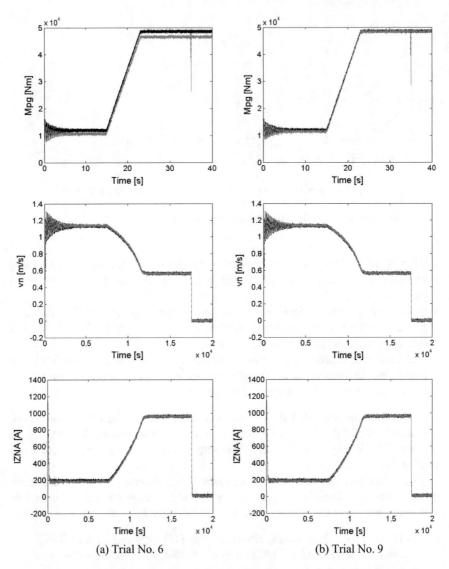

Fig. 4 Examples of simulation results obtained for different values of parameters determined by the evolutionary optimization algorithm in the 6th and 9th trial (y_r - *black line*, y_m - *red line*)

lower sampling rate and resolution for accurate tuning of a longwall scraper conveyor model. The sampling rate equals to 1 Hz with the resolution equals to 16bits can be enough for this kind of problems to have the mean error significantly less than 10% provided that it involves the required number of measuring sensors.

Taking into account overall results of the study presented in Table 2 it can be concluded that the most important parameter is the damping factor. Even a small change in the value of this parameter can have a strong influence on the results of the simulation. On the other hand, the value of the loss of the torques of flexible and hydrodynamic couplings has almost no effect on the simulation.

5 Conclusions

In this paper, the authors proposed and verified a heuristic optimization method for tuning values of parameters of longwall scraper conveyor model. In the considered model five of parameters were defined as uncertain, and the tuning problem was defined over these parameters. In turn, nine output parameters of the longwall scraper conveyor model were selected for tuning procedures. The tuning procedure was performed using four selected operational scenarios which were the most representative for the performed task.

As it was shown in the introduction section, this problem was not enough investigated before in known studies related to analytical modeling of mining conveyors. Following this, the authors have formulated the optimization problem and applied the evolutionary computation in order to find the final solution. Two kinds of the criterion function was proposed. The first one was based on Minkowski's distance, whilst the second was prepared by means of weighted sub-criteria such as the mean absolute percent error, the cross-correlation function and the aggregate value of the maximum absolute errors. The study was performed in three steps, where the first step was focused on optimal tuning of model parameters with an error minimization criterion, while the second and third steps were focused on estimation of appropriate parameters of an evolutionary algorithm, and quality of measurement signals and the number of sensors available for signal acquisition, respectively. It was shown that the second type of the criterion function should be used in order to obtain the smallest mean errors during searching optimal values of parameters. The validation tasks confirm that the proposed tuning method is characterized by the high precision of tuning of uncertain parameters of the model and has high potential from a practical point of view and what is more, it may be successfully applied in real mining conditions.

Acknowledgements The research presented in the paper was financed by the National Centre of Research and Development (Poland) within the framework of the project titled "An integrated shell decision support system for systems of monitoring processes, equipment and hazards" carried out in the path B of Applied Research Programme - grant No. PBS2/B9/20/2013. This publication is financed from the statutory funds of the Faculty of Mechanical Engineering of the Silesian University of Technology in 2016.

References

1. Czuba, W., Gospodarczyk, P., Kulinowski, P.: Application of the Discrete Element Method for simulation of haulage of excavated material by longwall scraper conveyor. Symulacja w Badaniach i Rozwoju **1**(3), 213–221 (2010). [in Polish]
2. Dolipski, M.: Dyn. Chain Conveyors. The Silesian University of Technology Press, Gliwice (1997). [in Polish]
3. Dolipski, M., Remiorz, E., Sobota, P.: Determination of dynamic loads of sprocket drum teeth and seats by means of a mathematical model of the longwall conveyor. Arch. Min. Sci. **57**(4), 1101–1119 (2012)
4. Dolipski, M., Cheluszka, P., Remiorz, E., Sobota, P.: Follow-up chain tension in an armoured face conveyor. Arch. Min. Sci. **60**(1), 25–38 (2015)
5. Mao, J., Shi, J., Zhang, D., Wei, X.: Dynamic modeling and simulation of heavy scraper conveyor. J. China Coal Soc. **33**(1), 103–106 (2008). [in Chinese]
6. Zhang, C., Meng, G.: Dynamic modeling of scraper conveyor sprocket transmission system and simulation analysis. In Proceedings of the 2011 IEEE International Conference on Mechatronics and Automation, pp. 1390–1394. Beijing (2011)
7. Eshin, E.K.: Modeling and management of dynamic state of conveyors. Vestnik KuzGTU **108**(2), 118–121 (2015). [in Russian]
8. Fan, Q., Wu, Y., Yu, Z.: Task coordination control modeling for coal machinery based on generalized partial global planning. J. Comput. Inf. Syst. **11**(2), 501–513 (2015)
9. Cenacewicz, K., Przystałka, P.: Conveyor belt simulator with fault models. Model. Eng. **24**(55), 13–20 (2015). [in Polish]
10. Cenacewicz, K., Katunin, A.: Modeling and simulation of longwall scraper conveyor considering operational faults. Studia Geotechnica et Mechanica **38**(2), 15–27 (2016)
11. Herbuś, K., Szewerda, K., Świder, J.: The concept of development and parametrisation of a virtual model of an armoured face conveyor (AFC). Model. Eng. **24**(55), 34–41 (2015). [in Polish]
12. Mucherino, A., Seref, O.: Advances in modeling agricultural systems, chapter. In: Modeling and Solving Real-Life Global Optimization Problems with Meta-heuristic Methods, pp. 403–419. Springer, Boston, MA, US (2009)
13. Vasant, P.M., Vasant, P.M.: Handbook of Research on Novel Soft Computing Intelligent Algorithms: Theory and Practical Applications, 1st edn. IGI Global, Hershey, PA, USA (2013)
14. Valadi, J., Siarry, P.: Applications of Metaheuristics in Process Engineering. Springer Publishing Company, Incorporated (2014)
15. Karolewski, B., Ligocki, P.: Modelling of long belt conveyors. Eksploatacja i Niezawodnosc - Maintenance and Reliability **16**(2), 179–187 (2014)
16. Deb, M.: Multi-Objective Optimization Using Evolutionary Algorithms. Wiley, New Jersey (2009)
17. Plamitzer, A.M.: Electrical Machines, 4th edn. WNT, Warsaw (1976). [in Polish]

Solving Graph Partitioning Problems with Parallel Metaheuristics

Zbigniew Kokosiński and Marcin Bała

Abstract In this article we describe computer experiments while testing a family of parallel and hybrid metaheuristics against a small set of graph partitioning problems like clustering, partitioning into cliques and coloring. In all cases the search space is composed of vertex partitions satisfying specific problem requirements. The solver application contains two sequential and nine parallel/hybrid algorithms developed on the basis of SA and TS metaheuristics. A number of tests are reported and conclusions concerning metaheuristics' performance that result from the conducted experiments are derived. The article provides a case study in which partitioning numbers $\psi_k(G)$, $k \geq 2$, of DIMACS graph coloring instances are evaluated experimentally by means of H-SP metaheuristic which is found to be the most efficient in terms of solution quality.

Keywords Simulated annealing · Tabu search · Parallel metaheuristic · Hybrid metaheuristic · Graph partitioning problem · Graph partitioning number

1 Introduction

Computational optimization attracts researchers and practitioners interested in solving combinatorial problems by means of various computational methods and tools. In particular, many NPO problems require new versatile tools in order to find approximate solutions [1, 10]. Parallel and hybrid metaheuristics are among the most promising methods to be developed in the nearest time [2, 20]. Many new algorithms have been already designed and compared with existing methodologies [7, 15], but there is still a room for significant progress in this area.

Z. Kokosiński (✉)
Faculty of Electrical and Computer Engineering, Cracow University of Technology,
ul. Warszawska 24, 31-155 Kraków, Poland
e-mail: zk@pk.edu.pl

M. Bała
Salumanus Sp. z o.o., ul. Walerego Sławka 8a, 30-633 Kraków, Poland
e-mail: bala.marcin@gmail.com

© Springer International Publishing AG 2018   89
S. Fidanova (ed.), *Recent Advances in Computational Optimization*,
Studies in Computational Intelligence 717, DOI 10.1007/978-3-319-59861-1_6

In this article we focus on a class of partitioning problems that appear in many application areas like data clustering [3], column-oriented database partitioning optimization [19], design of digital circuits, decomposition of large digital systems into a number of subsystems (models) for multi-chip implementation, task scheduling, timetabling, assignment of frequencies in telecommunication networks, etc. Partitioning problems are in general simpler than permutation problems but their search spaces are too huge for exhaustive search or extensive search methods [4–6, 11, 13, 21, 22].

The paper describes two computer experiments. The aim of the first one is determining of efficient metaheuristics for the given problem taking into account both quality of the solution (cost function) and the computation time. In many cases the tradeoff is not easy to find, similarly as the best algorithms' settings. However, from our research some general recommendations can be derived. In the second experiment a single algorithm with the best solution quality is used for finding solution of the Graph Partitioning Problem (GPP) for a class of graphs from DIMACS graph repository [8, 9]. Originally, they were used as instances of Graph Coloring Problem (GCP), which is known to be NP-complete. The obtained computational results provide additional characteristic of this collection of graphs, since the objective function used represents the cost of graph partitioning into exactly k clusters (partition blocks), $2 \leq k \leq 5$. The cost function minima found experimentally are the upper bounds for the partitioning numbers $\psi_k(G)$, which represent cost of the optimal clustering.

The rest of the paper is organized as follows. In the next section the graph partitioning problems are defined and characterized. Then, in Sect. 3, SA and TS algorithms as well as their parallelization and hybridization methods are sketched. The design assumptions and features of the developed solver application are described in Sect. 4. Testing methodology and initial experimental results are shown in Sect. 5. The experimental evaluation of $\psi_k(G)$ for 18 DIMACS graphs is described in detail in Sect. 6. The conclusions of the article point out the directions of future research in this area.

The presented research was conducted within the frame of statutory activity (grant No. E3/627/2016/DS) at Faculty of Electrical and Computer Engineering, Cracow University of Technology and was financially supported by Ministry of Science and Higher Education, Republic of Poland.

2 Graph Partitioning Problems

In this section formulations of several partitioning problems are given that are to be solved by a collection of algorithms used in the experimental part of the paper.

We assume that $G = (V, E)$ is a connected, undirected graph. Let $|V| = n$, $|E| = m$.

2.1 Graph Partitioning Problem (GPP)

A partition $C = (C_1, \ldots, C_k)$ of V is called a partitioning (clustering) of G and C_i clusters. C is called trivial if either $k = 1$, or all clusters C_i contain only one element. We will identify a cluster C_i with the induced subgraph of G, i.e. the graph $G_i = (C_i, E(C_i))$, where $E(C_i) = \{\{u, v\} \in E : u, v \in C_i\}$. Hence, $E(C) = \sum_{i=1}^{k} E(C_i)$ is the set of intra-cluster edges and $E \setminus E(C)$ the set of inter-cluster edges [3].

The number of intra-cluster edges is denoted by $m(C)$ and the number of inter-cluster edges by $M(C)$.

The *coverage*(C) of a graph clustering C is a fraction of intra-cluster edges within the complete set of edges E: $coverage(C) = m(C)/m$. The larger the value of *coverage*(C) does not necessarily mean the better quality of a clustering C.

Constructing a k-clustering with a fixed number of $k, k \geq 3$ of clusters is NP-hard [1].

In general, for k-clustering problems in weighted graphs the total weight of the set $E \setminus E(C)$ shall be minimized.

Unweighted graph instances G, like DIMACS graphs investigated in Sect. 6, can be characterized by the partitioning number $\psi_k(G)$ which equals to minimum $M(C)$.

2.2 Clique Partitioning Problem (CPP)

A partition $C = (C_1, \ldots, C_k)$ of V is called a partition of G into cliques iff every subgraph $G_i = (C_i, E(C_i))$ induced by a cluster C_i is a clique, i.e. all vertices in C_i are pairwise connected. The goal is to find the minimal k, for which a partition into at most k cliques exists.

The clique partitioning problem is NP-complete [18]. The dual problem to CPP is graph partitioning into independent sets (ISs). It is equivalent to the CPP for $G(V, E')$, where E' is a complement of the set E.

2.3 Clique Partitioning Problem with Minimum Clique Size (CPP)

In the present paper a solution of clique partitioning problem is also searched for given clique size at least s: is there a graph partition into k cliques satisfying a condition related to the minimum clique size s? For given n and k the minimum size of cliques in G is $s = \lfloor n/k \rfloor$. Weighted version of the problem are also known, with additional conditions related to cliques' weights [11].

2.4   Graph Coloring Problem (GCP)

Classical vertex coloring problem in a graphs is another formulation of graph partitioning into independent sets (ISs). Such ISs can be assigned different colors, satisfying the property that all pairs of adjacent vertices in G are assigned nonconflicting colors. Formally:

For given graph $G(V, E)$, the optimization problem GCP is formulated as follows: find the minimum positive integer $k, k \leq n$, and a function $c : V \longrightarrow \{1, \ldots, k\}$, such that $c(u) \neq c(v)$ whenever $(u, v) \in E$. The obtained value of k is referred to as graph chromatic number $\chi(G)$.

GCP belongs to the class of NP-complete problems [10].

2.5   Restricted Coloring Problem (RCP)

In practical applications a conflict-free vertex/edge coloring is searched, often satisfying additional requirements. Therefore, a large number of particular coloring problems arised and has been investigated [17].

One well known example is vertex coloring with some restrictions set on available colors for the given graph vertex. In RCP each vertex is assigned a list of forbidden colors and a proper solution meeting such set of constraints is searched [16].

3   Sequential and Parallel Metaheuristics

The reported research is based on two sequential and nine parallel algorithms. The sequential metaheuristics include classical simulated annealing (SA) and tabu search (TS) that belong to the class of iterative methods [20]. Parallel algorithms can be splitted into three categories: parallel metaheuristics derived from SA, parallel metaheuristics derived from TS and hybrid methods.

3.1   Simulated Annealing (SA)

Classical simulated annealing [20] is a well known technique widely used in optimization and present in most of the textbooks. It can be easily parallelized in various ways. Parallel moves enable single Markov chain to be evaluated by multiple processing units calculating possible moves from one state to another. Multiple threads compute independent chains of solutions and periodically exchange the obtained results. The key question in parallel implementation remains setting of algorithm's parameters like initial temperature, and a cooling schedule. For the problem at hand it is necessary to define an appropriate solution representation, cost function and a neighborhood generation scheme.

3.2 Tabu Search (TS)

Tabu search [20] is an improvement of local search method in which so called tabu list contains a number of recent moves that must not be considered as candidates in the present iteration. This feature helps the method to escape from local minima what is impossible in local search. The question is to define the solution representation, cost function, neighborhood and a single move, the size of the neighborhood and the number of candidate moves, aspiration level which decides on the possibility to accept forbidden moves if it leads to a solution improvement etc.

3.3 MIR Model of Parallelization

Multiple independent runs (MIR) model is a very popular way of parallelization of iterative algorithms. A number of algorithm instances with different input data are executed simultaneously. All computational processes run independently and do not exchange data during computation. At the end, the best solution from all processes is selected. This simple model can be made more sophisticated by introducing an information exchange scheme, exchange rate etc.

3.4 MS Model of Parallelization

In Master-Slave (MS) model the master executes the sequential part of an algorithm, distributes computational tasks among slaves, collects results from slaves, process and aggregates this results. In certain versions of MS model the master splits the whole search space among slaves, synchronizes their work, checks the termination condition and collects the best solution from subspaces.

3.5 PA Model of Parallelization

Parallel asynchronous (PA) model provides maximum flexibility: various algorithms with different initial data search the whole search space in an asynchronous manner. Usually an efficient update scheme for the best solution must be implemented as well as occasional distribution of best solutions to asynchronous computational processes. One possibility is to employ a communication process. In some cases shared memory (SM) can be used for information updates and exchange. The second solution helps to avoid generation of interrupts in asynchronous processes. The processes communicate the SM in predictable moments of time.

3.6   Hybrid Models

Hybrids models include: 1. two-phase algorithms, when each phase - restriction of the search space and solution refinement - is performed by a different method; 2. combined algorithms, when known elements of existing methods are composed in a single algorithm; 3. combined algorithms consisting original components like problem-oriented operations or heuristics; and 4. concurrent algorithm which is parallel execution of known methods with data exchange patterns.

In this paper three parallel metaheuristic algorithms are used.

Parallel hybrid asynchronous (H-PA) algorithm splits computational processes into "even" performing SA and "odd" performing TS. Best solutions are updated via shared memory SM, where they are immediately made available for all processes.

Hybrid serial-parallel algorithm (H-SP) process in parallel p threads in which SA and TS sections are performed alternately starting from SA section. SA section modifies tabu list while TS section modifies current temperature for the next section, respectively. Switching conditions are related to the progress achieved in improving best solution.

Parallel hybrid algorithm (H-P) is developed on the basis MIR method. Single step combines properties of both SA and TS: if new solution satisfies aspiration criterion (AC) it is always accepted, otherwise, it is accepted according to SA rules. This means that probability of acceptance of worse solution decreases in time.

4   The Solver

For all tests the "Partitioning problems solver" application was used. It was written in C++ (Visual Studio), while .NET Framework 3.5 provided necessary libraries and runtime environment.

The main program window contains three tabs: Program, Generator and Help. In appropriate fields of Program tab it is possible to select one of five basic problems (GPP, CPP, CPP-MIN, GCP, RGCP) and one of eleven algorithms. After that, one can select the input file format and read input data. A numerous algorithm parameters and problem constrains must be filled in the forms including multiple runs, enabling statistics and write options. The cost of best solution and the total computation time are also displayed in this tab.

The Generator tab opens possibilities to generate input graphs or weighted input graphs after setting its parameters and lists of forbidden colors. The unweighted graphs are kept in .col format, weighted graphs are in .ecl format, which is extension of .col by adding edge weights as well as edge weight range (in the header). The type .rcp contains lists of forbidden colors for all vertices, if any. File formats .xpp and .xcp are used for preserving input graph and the partition being the best solution for the given problem together with its cost, respectively. Output data in CSV format are written into the .txt file and enable easy import of data to a spreadsheet.

5 Computational Experiments with Metaheuristics

For experiments Intel Pentium T2300 machine was used with two 1.66 GHz cores and 4GB RAM, running under Windows XP Pro SP2 and .NET Framework 3.5 platform.

All five problems were tested against all eleven algorithms with eight basic settings (stop criterion, no of iterations in a single step, initial temperature for SA, size of the tabu list). The specific setting that were selected in the initial phase of the experiment are shown in Table 1.

Other parameters are: coefficient of cost function = 1, no of parallel processes (threads) = 20, communication parameter = 20, no of algorithm repetitions = 20, no of clique extension trials = 5, no of repetitions for H-SP algorithm = 5.

In Tables 2, 3, 4, 5, 6, 7, 8, 9, 10 and 11 computational data are presented. All experiments were conducted for random graph instances generated for each class of the graph partitioning problems in .ecl format. Relatively small graph instances were used with 20, 50 and 100 vertices and graph densities 10, 20 and 30%. Cost functions from 20 trials are collected in Tables 2, 3, 4, 5 and 6 while the corresponding computation times in Tables 7, 8, 9, 10 and 11, respectively.

Analysis of the results obtained for the five partitioning problems justifies several conclusions.

The shortest processing times are obtained by pure TS and SA methods. However, their solutions are not satisfactory. Parallelization and hybridization require additional computational work, and their aim is to improve search for a better suboptimal solution rather then providing significant speedup.

For GPP the fastest parallel algorithms is P-TS metaheuristic. PSA and two hybrid methods (H-PA, H-P) are less time-efficient. The slowest algorithm is H-SP, which is very time consuming. On the other hand H-SP finds the best solutions for all eight available settings. Average results of SA-MIR and H-P algorithms are also outstanding and obtained approximately five times faster than by H-SP. The best setting in average is no 6 (minimum cost for six methods), but the best result for

Table 1 Basic settings of algorithms

No.	Stop criterion (it)	Number of iterations/step	SA - initial temperature	TS - size of tabu list
1	20	5	3	10
2	20	5	10	40
3	20	10	3	10
4	20	10	10	40
5	50	5	3	10
6	50	5	10	40
7	50	5	3	10
8	50	5	10	40

Table 2 Graph partitioning problem (GPP). Cost functions (11 algorithms, 8 settings, 20 runs)

	SA	PSA			TS	PTS			Hybrid			Avg.
		MIR	MS	A		MIR	MS	A	H-PA	H-SP	H-P	
1	295	238	255	251	338	284	285	287	286	230	240	271.7
2	296	237	254	253	330	286	285	282	246	228	234	266.5
3	293	243	257	259	338	283	285	288	254	234	238	270.2
4	293	238	257	257	331	282	285	283	252	232	235	267.7
5	292	234	249	247	341	276	286	288	245	226	237	265.5
6	286	235	245	244	338	274	284	281	238	227	235	262.5
7	289	242	252	256	332	280	286	286	249	238	242	268.4
8	289	240	252	252	329	271	280	283	252	235	239	265.6
Avg.	291.6	238.4	252.6	252.4	334.6	279.5	284.5	284.8	252.8	**231.2**	237.5	

Table 3 Clique partitioning problem (CPP). Cost functions (11 algorithms, 8 settings, 20 runs)

	SA	PSA			TS	PTS			Hybrid			Avg.
		MIR	MS	A		MIR	MS	A	H-PA	H-SP	H-P	
1	26	24	24	24	25	23	24	24	24	21	24	23.9
2	26	24	24	24	25	23	24	24	24	21	24	23.9
3	23	22	22	22	24	22	23	23	23	21	22	22.5
4	24	22	22	22	24	22	23	23	23	21	22	22.5
5	25	24	24	24	24	22	23	23	23	21	24	23.4
6	26	24	24	24	24	22	23	23	23	21	24	23.5
7	23	22	22	22	23	22	22	22	22	21	22	22.1
8	24	22	22	22	23	22	22	22	22	21	22	22.2
Avg.	24.6	23	23	23	24	22.3	23	23	23	**21**	23	

Table 4 CPP with min. clique size (CPP-MIN). Cost functions $\times 10^3$ (11 algorithms, 8 settings, 20 runs)

	SA	PSA			TS	PTS			Hybrid			Avg.
		MIR	MS	A		MIR	MS	A	H-PA	H-SP	H-P	
1	114	107	108	108	147	126	135	134	134	101	106	120
2	115	107	108	108	144	127	135	134	110	101	106	118
3	111	105	105	105	137	122	129	129	108	100	104	114
4	112	105	105	105	137	123	129	128	107	100	105	114
5	112	105	106	106	140	114	131	130	108	101	106	114
6	113	106	106	106	139	114	130	130	108	101	106	114
7	112	105	105	105	133	113	126	126	107	99	105	112
8	112	104	105	105	132	112	126	126	107	99	104	112
Avg.	113	105	106	106	139	119	130	130	111	**100**	105	

Table 5 Graph coloring problem (GCP). Cost functions (11 algorithms, 8 settings, 20 runs)

	SA	PSA			TS	PTS			Hybrid			Avg.
		MIR	MS	A		MIR	MS	A	H-PA	H-SP	H-P	
1	86	51	52	53	84	55	55	55	55	55	54	59.5
2	80	52	52	52	85	55	54	55	53	53	53	58.5
3	85	52	51	52	83	54	55	54	54	54	53	58.8
4	86	51	52	51	78	53	53	53	53	53	53	57.8
5	82	52	52	52	86	54	56	55	54	54	53	59.1
6	87	52	52	52	86	53	54	55	54	53	53	59.2
7	84	50	51	52	83	53	54	55	54	53	53	58.4
8	86	52	51	52	85	53	54	53	52	53	53	58.5
Avg.	84.5	**51.5**	51.6	52	83.8	53.8	54.4	54.4	53.6	53.5	53.1	

Table 6 Restricted GCP (RGCP). Cost functions (11 algorithms, 8 settings, 20 runs)

	SA	PSA			TS	PTS			Hybrid			Avg.
		MIR	MS	A		MIR	MS	A	H-PA	H-SP	H-P	
1	36	26	26	27	39	29	30	29	30	27	29	29.8
2	37	26	26	26	37	28	29	29	27	26	28	29
3	36	26	26	27	39	28	29	28	28	26	28	29.2
4	36	27	27	27	36	28	28	28	27	26	28	28.9
5	37	26	26	26	38	28	29	29	27	26	28	29.1
6	36	26	27	26	38	27	28	29	27	27	28	29
7	36	26	27	27	37	28	28	28	27	26	28	28.9
8	36	26	27	27	36	27	28	28	27	26	27	28.6
Avg.	36.3	**26.1**	26.5	26.6	37.5	27.9	28.6	28.5	27.5	26.3	28	

Table 7 Graph partitioning problem (GPP). Computation times (11 algorithms, 8 settings, 20 runs)

	SA	PSA			TS	PTS			Hybrid			Avg.
		MIR	MS	A		MIR	MS	A	H-PA	H-SP	H-P	
1	1.70	19.4	14.6	15.3	0.57	6.83	5.23	5.13	5.49	70.0	34.2	16.2
2	1.76	18.3	13.4	14.2	0.60	6.75	5.40	5.51	14.1	67.5	18.4	15.1
3	6.28	64.2	48.4	50.0	2.48	26.5	21.9	21.3	40.1	27.0	64.9	56.0
4	5.71	58.3	42.8	41.5	2.85	29.8	26.9	25.6	40.1	258	58.7	53.7
5	3.77	35.0	32.2	30.4	1.5	25.3	13.1	13.5	27.8	149	35.4	33.3
6	3.33	34.4	31.3	32.4	1.76	33.4	15.1	15.0	28.2	149	34.9	34.4
7	10.7	106	97.7	98.1	6.59	114	55.4	54.8	85.9	626	109	124
8	10.2	101	91.7	92.5	6.98	128	67.1	65.6	86.5	605	104	123
Avg.	5.4	54.8	46.5	46.8	2.92	46.3	26.3	25.8	41.0	274	57.5	

Table 8   Clique partitioning problem (CPP). Computation times (11 algorithms, 8 settings, 20 runs)

	SA	PSA			TS	PTS			Hybrid			Avg.
		MIR	MS	A		MIR	MS	A	H-PA	H-SP	H-P	
1	1.03	9.86	10.0	9.82	1.05	10.8	10.0	10.5	10.5	76.4	10.0	14.5
2	0.95	9.46	9.59	9.40	1.06	11.0	10.5	10.4	10.2	76.6	9.62	14.4
3	3.45	33.9	33.3	32.7	3.30	34.0	31.8	31.5	32.7	280	33.7	50.1
4	3.09	30.9	30.5	30.8	3.33	33.3	31.6	31.9	32.6	282	31.1	49.2
5	1.89	18.2	18.3	18.4	2.14	23.5	21.5	21.7	22.1	168	18.2	30.4
6	1.78	17.6	17.9	17.9	2.27	23.5	22.0	21.6	21.2	158	17.4	29.2
7	6.50	63.8	63.8	63.5	7.25	77.9	70.2	71.4	73.5	653	64.8	110
8	6.10	61.8	60.9	60.9	7.30	76.3	72.5	70.2	72.4	645	61.2	109
Avg.	3.10	30.7	30.5	30.4	3.46	36.3	33.8	33.7	34.4	292	30.8	

Table 9   CPP with min. clique size (CPP-MIN). Computation times (11 algorithms, 8 settings, 20 runs)

	SA	PSA			TS	PTS			Hybrid			Avg.
		MIR	MS	A		MIR	MS	A	H-PA	H-SP	H-P	
1	1.74	17.6	16.3	17.2	1.25	18.1	13.0	13.6	13.3	91.2	18.0	20.1
2	1.76	17.0	16.7	17.3	1.28	17.5	13.4	13.6	16.6	89.4	16.6	20.1
3	3.89	39.9	39.6	39.4	3.82	53.3	40.5	39.3	40.0	337	39.7	61.6
4	3.90	38.4	38.2	39.2	3.90	52.1	39.4	39.5	40.1	332	38.4	60.5
5	2.93	27.6	29.6	29.3	2.60	50.5	26.7	26.4	30.0	193	26.9	40.5
6	2.85	27.4	29.9	29.3	2.58	49.6	27.1	27.3	29.7	185	27.4	39.9
7	7.24	72.3	72.1	73.4	8.60	169	89.1	83.0	77.6	795	71.9	138
8	7.33	71.9	73.1	71.8	9.04	167	90.4	86.6	76.6	840	71.9	142
Avg.	3.96	39.0	39.4	39.6	41.4	72.2	42.4	41.2	40.5	358	38.9	

Table 10   Graph coloring problem (GCP). Computation times (11 algorithms, 8 settings, 20 runs)

	SA	PSA			TS	PTS			Hybrid			Avg.
		MIR	MS	A		MIR	MS	A	H-PA	H-SP	H-P	
1	0.62	7.09	7.14	6.97	0.64	7.06	7.01	6.88	6.84	49.3	8.13	9.79
2	0.60	6.56	6.57	6.63	0.68	8.05	7.68	7.35	6.36	48.9	8.08	9.77
3	2.36	26.9	26.2	26.2	2.38	27.2	24.6	25.2	23.3	201	27.9	37.6
4	2.24	25.1	24.6	24.9	2.31	27.1	26.0	26.1	22.4	201	27.5	37.2
5	1.37	14.4	14.4	14.3	1.51	17.9	15.9	15.6	15.6	121	18.3	22.8
6	1.31	13.9	14.0	14.0	1.45	19.2	17.6	18.0	20.4	122	18.9	23.7
7	5.35	56.9	56.4	56.1	5.73	65.1	61.3	60.0	55.8	498	68.3	90.0
8	5.24	54.5	55.3	54.7	5.37	61.3	56.3	56.8	52.8	499	59.6	87.4
Avg.	2.39	25.7	25.6	25.5	2.51	29.1	27.1	27.0	25.4	218	29.6	

Table 11 Restricted GCP (RGCP). Computation times (11 algorithms, 8 settings, 20 runs)

	SA	PSA			TS	PTS			Hybrid			Avg.
		MIR	MS	A		MIR	MS	A	H-PA	H-SP	H-P	
1	0.77	7.73	7.58	7.92	0.66	7.58	6.56	6.61	6.59	49.0	7.35	9.85
2	0.72	7.15	7.25	7.07	0.71	7.80	7.35	7.24	6.20	49.1	8.01	9.87
3	2.78	28.7	2.89	29.4	2.51	27.1	25.7	25.2	23.1	199	28.1	38.3
4	2.64	26.9	26.7	26.7	2.57	28.7	26.9	26.2	22.9	199	28.9	38.0
5	1.51	14.9	15.2	15.3	1.58	18.3	15.6	16.1	14.1	121	18.4	22.9
6	1.45	14.3	14.5	14.6	1.68	19.2	17.5	16.5	14.6	121	19.2	23.2
7	5.84	57.9	58.6	58.7	6.26	72.3	63.6	62.7	55.0	497	72.4	91.8
8	5.61	56.4	56.1	56.7	5.79	66.0	63.6	61.9	55.3	496	66.6	90.0
Avg.	2.66	26.8	26.9	27.1	2.72	30.9	28.3	27.8	24.7	217	31.1	

GPP is obtained with setting no 5. In terms of the computation time settings no 2 and 1 obviously win, and the fastest method is the TS-PA algorithm with moderate success in optimization.

For CPP the fastest parallel algorithms are PSA metaheuristics. Two hybrid methods (H-PA, H-P) are also timeefficient. Among the parallel algorithm SA-PA is the fastest one with minimum obtained for four settings. The slowest algorithm is again H-SP, which finds the best solutions for all eight parameter settings. The second result provides TS-MIR which is eight times faster than H-SP. Setting no 7 provides the best solution quality for all algorithms. In terms of the computation time settings no 2 and 1 win, and the fastest method is the SA-PA algorithm.

For CPP-MIN the fastest parallel algorithms are hybrid and PSA metaheuristics. The winner is H-P algorithm with setting no 2. The slowest algorithm is H-SP, which wins the quality competition for all eight parameter settings. SA-MIR and H-P has been the most prospective challengers. Setting no 8 provides the best solution quality for nine algorithms. In terms of the computation time settings no 1 and 2 are the winners.

For GCP the fastest parallel methods are PSAs which provide also best approximate solutions (SA-MIR wins for six out of eight settings). The fastest parallel algorithm is H-PA, the best setting for seven algorithms is no 2. The slowest algorithm is H-SP, which is 4th in terms of solution quality. The best setting for cost-optimality is no 4 in average.

The last problem - RGCP - brings also interesting results. The fastest parallel algorithms are PSAs, but the winner in this category is H-PA with seven winning settings. The best settings for all methods are 1 and 2. The best solution in average is found by PSA-MIR (the winner for seven out of eight settings), the runner-up is H-SP which was about eight time slower, the next positions are occupied by PSA-MS and PSA-A. Most good results (9) were obtained for the setting no 8.

6   Graph Partitioning Problem - A Case Study

Graphs are often characterized by their combinatorial properties. For instance, in the second DIMACS Implementation Challenge: 1992–1993, chromatic numbers $\chi(G)$ were searched for a collection of hard to color graph instances by means of virtually all known computational techniques. Gradually, most graph instances from this repository were assigned these distinctive numbers. The first parallel metaheuristic used for searching chromatic numbers $\chi(G)$ was Parallel Evolutionary Algorithm (PEA) [15]. Another example of graph characteristics are graph chromatic sum $\sum(G)$ and graph chromatic sum number $s(G)$ [17]. In the research on sum coloring PEA as well as many new computational methods were highly successful [12, 14].

We believe that DIMACS graphs could be further characterized with respect to other graph partitioning problems. In this context GPP seems to be one of the most important and promising candidates. In the next experiment we will investigate 18 DIMACS graphs in search for their clustering numbers $\psi_k(G)$, for different number of clusters $k \geq 2$. For our computations hybrid serial–parallel algorithm (H-SP) has been selected due to its efficiency in solving majority of partitioning problems. The time efficiency is not a priority, but the number of algorithm's runs shall be reasonably restricted.

The algorithm H-SP uses p threads, each with a pair of modified SA and TS algorithms. SA section creates a *tabu list* for TS on exit while TS section updates *temperature* for SA on exit. Starting from SA section, both H-SP components work by turns improving the current and the best solutions, respectively. Alternate runs are continued till the stop criterion is reached. Details of SA and TS algorithms are omitted here.

The pseudocode of the H-SP algorithm used in our experiment is as follows.

```
H-SP(input: p,stop_criterion,iter_number,initial_temp,
list_size, alternat_coeff; output: best_sol,cost(best_sol));
best_sol(1):= random_vertex_partition(1)
best_sol:= best_sol(1)
cost_best_sol:=cost(best_sol(1))
for j=1 to p do in parallel
current_sol(j):= random_vertex_partition(j)
best_sol(j):=current_sol(j)
tabu_list(j):=empty
T(j):=initial_temp
repeat
  alternat_counter:=0
  repeat
    for iter_counter=1 to iter_number do
      SA(T(j),temp_coeff,current_sol(j),best_sol(j));
      if (cost(best_sol(j)) < cost_best_sol)
        then best_sol:=best_sol(j)
```

```
                cost_best_sol:=cost(best_sol(j))
                iter_counter:=0
          else Inc(alternat_counter)
   until (alternat_counter > alternat_coeff)
   update(tabu_list(j))
   alternat_counter:=0
   repeat
     for iter_counter=1 to iter_number do
       TS(tabu_list(j),list_size,current_sol(j),best_sol(j));
       if (cost(best_sol(j)) < cost_best_sol)
         then best_sol:=best_sol(j)
              cost_best_sol:=cost(best_sol(j))
           iter_counter:=0
         else Inc(alternat_counter)
   until (alternat_counter > alternat_coeff)
   update(T(j))
   Inc(main_counter)
until (main_counter = stop_criterion)
```

For H-SP algorithm the setting no. 2 from Table 1 was chosen what is justified by results of the reported testing. Other program settings for H-SP were as follows: *no. of repetitions* = 10, *no. of parallel processes* = 20, *no. of algorithm's runs* = 10, 50.

For this experiment the machine with Intel Core i7 4700MQ CPU was used with four 2.4 GHz cores and 8 GB RAM, running under Windows 10 Home OS.

The results of computations are shown in Tables 12 and 13. In Table 12 nine DIMACS graphs are gathered (5 book graphs, *games120* and 3 *miles* graphs). Due to excessive time *homer* graph was processed only with 10 algorithm's runs. Table 13 contains nine *queen* graphs. All computed upper bounds for $\psi_k(G)$, $2 \le k \le 5$, are distinguished in a bold font. Let us notice, that in several cases the bounds obtained with 10 algorithm's runs are better than with 50 runs. The presented computational results should be considered as the first attempt in evaluation of numbers $\psi_k(G)$ for DIMACS graph coloring benchmarks.

7 Conclusions

In this article some research results related to parallel metaheuristics and their applications were reported. The conducted experiments gave certain limited insight to computational behaviour of parallel and hybrid metaheuristics developed on the basis of SA and TS algorithms, and applied to a class of popular partitioning problems in graphs. Some algorithms were better than others for solving particular problems. We were focused mostly on solution quality, the computation time was the secondary factor in our comparison. Many obtained results were not obvious and difficult to predict without experimental verification.

Table 12 Evaluation of $\psi_k(G)$ for GPP (DIMACS graphs, algorithm H-SP, *no. of runs* = 10, 50)

Graph	No of blocks	10 runs			50 runs		
		Best cost	Iterations best/total	Best run (s)	Best cost	Iterations best/total	Best run (s)
anna	2	**139**	28056/75840	1.703	142	85440/133200	1.703
	3	151	56232/104160	2.343	**148**	104448/152400	2.828
	4	158	62832/110640	2.421	**150**	98088/145920	3.185
	5	157	99048/146880	3.156	**155**	87840/135600	3.015
david	2	141	39696/231360	2.468	**132**	80352/272160	2.703
	3	142	251712/443520	4.390	**136**	65976/257760	2.546
	4	148	205152/396960	3.937	**147**	55368/247200	2.468
	5	149	184776/376320	4.031	**147**	104976/296640	3.093
homer	2	**445**	576456/768000	278.6	n.a.	n.a.	n.a.
	3	**420**	356232/548160	187.2	n.a.	n.a.	n.a.
	4	**434**	439152/630720	215.9	n.a.	n.a.	n.a.
	5	**444**	453888/645600	219.7	n.a.	n.a.	n.a.
huck	2	61	56328/248160	1.765	**59**	101832/293760	2.078
	3	64	17232/208800	1.468	**63**	23880/215520	1.578
	4	69	139176/330720	2.625	**67**	193632/385440	2.765
	5	**72**	339000/530880	3.937	**72**	31440/223200	1.546
jean	2	52	188664/380640	3.125	**51**	38136/229920	2.031
	3	**53**	176856/368640	3.035	54	113400/305280	2.734
	4	60	240264/432000	3.890	**59**	50976/242880	2.187
	5	62	48096/230400	2.031	**61**	49320/240960	2.140
games120	2	133	65064/256800	4.640	**128**	48600/240480	4.641
	3	128	461424/653280	12.04	**123**	86256/277920	5.359
	4	**127**	282792/474720	8.718	136	301992/493920	9.594
	5	149	47592/239520	4.359	**138**	218184/409920	7.968
miles250	2	33	314592/506400	9.484	**29**	290568/482400	9.797
	3	34	59064/251040	4.859	**30**	307944/499680	9.859
	4	**35**	77136/268800	5.000	37	73968/265920	5.219
	5	**43**	92400/284160	5.250	46	75048/266880	5.770
miles500	2	159	89256/280800	6.906	**155**	39072/230880	5.826
	3	160	84504/276480	6.468	**156**	54696/246240	6.209
	4	171	256104/447840	10.93	**163**	128904/320640	8.125
	5	168	63960/255840	6.140	**167**	48792/240480	6.015
miles750	2	534	73776/265440	9.328	**524**	469200/660960	23.56
	3	529	278544/470400	16.82	**475**	101064/292800	10.56
	4	534	89040/280800	9.578	**530**	614904/806880	28.28
	5	539	132648/324480	11.59	**535**	338136/529920	18.26

Table 13 Evaluation of $\psi_k(G)$ for GPP (DIMACS graphs, algorithm H-SP, *no. of runs* = 10, 50)

Graph	No of blocks	10 runs			50 runs		
		Best cost	Iteration best/total	Best run (s)	Best cost	Iteration best/total	Best run (s)
queen5.5	2	81	34080/225600	0.531	**74**	25704/217440	0.375
	3	**80**	41640/233280	0.546	**80**	31272/223200	0.375
	4	102	10512/202080	0.484	**90**	42456/234240	0.390
	5	102	27240/218880	0.515	**92**	4536/196320	0.343
queen6.6	2	141	15096/206880	0.703	**137**	54360/246240	0.843
	3	155	20496/212160	0.734	**149**	20736/212640	0.578
	4	158	106032/297600	1.062	**148**	71328/263040	0.718
	5	164	36264/228000	0.609	**157**	104976/296640	0.999
queen7.7	2	249	32016/223680	1.078	**238**	32256/224160	0.937
	3	247	35232/227040	1.093	**237**	164400/356160	1.484
	3	251	278160/469920	2.328	**244**	29088/220800	1.078
	4	253	22104/214080	0.890	**244**	16488/208320	1.046
queen8.8	2	373	88176/279840	1.984	**338**	125568/317280	2.218
	3	377	45048/236640	1.640	**360**	27360/218880	1.546
	4	369	296928/488640	3.531	**353**	102648/294240	2.062
	5	384	273576/465120	3.281	**376**	354576/546240	3.453
queen8.12	2	665	263808/455520	5.968	**664**	90360/282240	4.359
	3	**667**	130656/322560	5.164	670	219768/411360	6.562
	4	**670**	180264/372000	5.945	674	379728/571680	9.156
	5	687	103008/294720	4.709	**678**	47280/239040	3.781
queen9.9	2	532	339816/531360	5.562	**521**	62832/254400	2.859
	3	**498**	44280/236160	2.421	531	53304/245280	2.734
	4	540	217584/409440	4.250	**537**	60216/252000	2.609
	5	**533**	229440/420960	4.812	**533**	63192/254880	2.671
queen10.10	2	**725**	64512/256320	3.562	**725**	239352/431040	6.062
	3	735	51552/243360	3.718	**726**	39816/231360	3.265
	4	**730**	141360/333120	4.986	**730**	606768/798720	11.34
	5	738	139296/331200	4.671	**734**	48384/240000	3.531
queen11.11	2	974	137496/329280	6.984	**962**	244488/436320	9.437
	3	970	58272/250080	5.460	**967**	78384/270240	6.031
	4	**971**	105000/296640	6.459	**971**	126456/318240	6.750
	5	981	56808/248640	5.577	**974**	59328/251040	5.344
queen12.12	2	1254	258216/354000	11.36	**1242**	125544/221520	6.750
	3	1249	37416/133200	4.296	**1241**	264984/360960	11.12
	4	1247	187992/283920	8.750	**1236**	335544/431520	13.71
	5	1270	31128/126960	3.828	**1250**	233064/329040	10.09

An original contribution of our research is evaluation of graph partitioning numbers $\psi_k(G)$, $2 \leq k \leq 5$, for 18 DIMACS graph coloring instances, which were so far characterized by chromatic number $\chi(G)$, chromatic sum $\sum(G)$ and graph chromatic sum number $s(G)$. Now, they have got also 72 experimentally computed upper bounds for $\psi_k(G)$.

We believe that the presented results justify further experiments with our solver. It would be interesting to extend our second experiment and include PSA-MIR algorithm which also provides quality results in a reasonable computation time. Another research direction is to continue evaluation of graph partitioning numbers for the remaining DIMACS graphs.

References

1. Ausiello, G., et al.: Complexity and Approximation – Combinatorial Optimization Problems and Their Approximability Properties. Springer, Berlin (1999)
2. Blum, C., Roli, A., Alba, E.: An introduction to metaheuristic techniques. In: Alba, E., et al. (eds.) Parallel Metaheuristics, pp. 3–42. Wiley-Interscience, New York (2005)
3. Brandes, U., Gaertler, M., Wagner, D.: Experiments on graph clustering algorithms. Lect. Notes Comp. Sci. (2003). doi:10.1007/978-3-540-39658-1-52
4. Byun, C.-Y.: Lower bound for large-scale set partitioning problems. ZIB-Report **12**, 18–22 (2001)
5. Charon, I., Hudry, O.: Noising methods for a clique partitioning problem. Discret. Appl. Math. (2006). doi:10.1016/j.dam.2005.05.029
6. Coslovich, L., Pesenti, R., Ukovich, W.: Large-scale set partitioning problems. J. Comput. **13**, 191–209 (2001)
7. Crawford, B., Castro, C.: ACO with Lookahead Procedures for Solving Set Partitioning and Covering Problems. Proc. of workshop on combination of metaheuristic and local search with constraint programming techniques, Nantes, France (2005)
8. COLOR web site. http://mat.gsia.cmu.edu/COLOR/instances.html
9. DIMACS ftp site. ftp://dimacs.rutgers.edu/pub/challenge/graph/benchmarks/
10. Garey, R., Johnson, D.S.: Computers and Intractability. A Guide to the Theory of NP-Completeness. Freeman, San Francisco (1979)
11. Ji, X., Mitchell, J.E.: The clique partition problem with minimum clique size requirement. Discret. Optim. **4**, 87–102 (2007)
12. Jin, Y., Hamiez, J.-P., Hao, J.-K.: Algorithms for the minimum sum coloring problem: a review. Artif. Intell. Rev. **47**, 367–394 (2017). doi:10.1007/s10462-016-9485-7
13. Kernighan, B.W., Lin, S.: An efficient heuristics procedure for partitioning graphs. Bell Syst. Tech. J. **49**, 291–307 (1970)
14. Kokosiński, Z., Kwarciany, K.: On sum coloring of graphs with parallel genetic algorithms. Lect. Notes Comp. Sci. (2007). doi:10.1007/978-3-540-71618-1_24
15. Kokosiński, Z., Kwarciany, K., Kołodziej, M.: Efficient graph coloring with parallel genetic algorithms. Comput. Inform. **24**, 109–121 (2005)
16. Kubale, M.: Some results concerning the complexity of restricted colorings of graphs. Discret. Appl. Math. **36**, 35–46 (1992)
17. Kubale, M. (ed.): Graph Colorings. American Mathematical Society (2004)
18. Mujuni, E., Rosamond, F.: Parameterized complexity of the clique partition problem. In: The Australasian Theory Symposium, pp. 75–78 (2008)
19. Nowosielski, A., Kowalski, P.A., Kulczycki, P.: The column-oriented database partitioning optimization based on the natural computing algorithms. Proc. FedCSIS (2015). doi:10.15439/2015F262

20. Sait, S.M., Youssef, H.: Iterative Computer Algorithms with Applications in Engineering. IEEE Computer Society, Los Alamitos (1999)
21. Tseng, W.-D., Hwang, I.-S., Lee, L.-J., Yang, C.-Z.: Clique-partitioning connections-scheduling with faulty switches in dilated Benes network. J. Chin. Inst. Eng. **32**, 853–860 (2009)
22. Zhou, S.: Minimum partition of an independence system into independent sets. Discret. Optim. (2009). doi:10.1016/j.disopt.2008.10.001

Comparison of Selected Fuzzy PSO Algorithms

Tomasz Krzeszowski, Krzysztof Wiktorowicz and Krzysztof Przednowek

Abstract This paper presents a comparison of selected fuzzy particle swarm optimization algorithms. Two non-fuzzy and four fuzzy algorithms are considered. The Takagi–Sugeno fuzzy system is used to change the parameters of these algorithms. A modified fuzzy particle swarm optimization method is proposed in which each of the particles has its own inertia weight and coefficients of the cognitive and social components. The evaluation is based on the common nonlinear benchmark functions frequently used for testing optimization methods. The ratings of the algorithms are assigned on the basis of the mean of the objective function final value.

1 Introduction

Particle swarm optimization (PSO) is a stochastic optimization method that was developed by Kennedy and Eberhart [10]. The PSO is mainly inspired by the social behavior of organisms that live and interact within large groups, for example, schools of fish, flocks of birds or swarms of bees. The usefulness of PSO in solving a wide range of optimization problems has been repeatedly confirmed. It has been applied to: the intelligent identification and control of a dynamic system [3]; human motion tracking [20]; solving an economic dispatch problem in power systems [19]; feature selection [2]; automatic incident detection [22]; the open shortest path first weight setting problem [17]; fuzzy anomaly detection in networks [9]; the estimation of hurdles clearance parameters [12]; clustering [7] and many more problems. Many

T. Krzeszowski (✉) · K. Wiktorowicz
Faculty of Electrical and Computer Engineering, Rzeszow University of Technology,
al. Powstańców Warszawy 12, 35-959 Rzeszów, Poland
e-mail: tkrzeszo@prz.edu.pl

K. Wiktorowicz
e-mail: kwiktor@prz.edu.pl

K. Przednowek
Faculty of Physical Education, University of Rzeszow,
ul. Towarnickiego 3, 35-959 Rzeszów, Poland
e-mail: krzprz@ur.edu.pl

© Springer International Publishing AG 2018   107
S. Fidanova (ed.), *Recent Advances in Computational Optimization*,
Studies in Computational Intelligence 717, DOI 10.1007/978-3-319-59861-1_7

modifications of the PSO have been developed since it was introduced in 1995 [10]. The most common are algorithms with a constriction factor [4] and with a linear inertia weight [5]. Among the PSO modifications we can distinguish algorithms that utilize fuzzy systems [1, 3, 8, 14, 15, 18, 19, 21, 24]. For example, in papers [3, 21] a fuzzy system was used to dynamically modify the inertia weight. In the paper [14], the authors present a fuzzy PSO with a cross-mutated operation (FPSOCM), where a fuzzy logic determines the inertia weight and the control parameter of the proposed cross-mutated operation. Another approach was presented in [19], where a fuzzy system is used to change the inertia weight and the coefficients of the cognitive and social components.

The main contributions of this paper are summarized as follows:

- we propose the MFPSO algorithm, in which each of the particles has its own inertia weight and the coefficients of the cognitive and social components,
- we compare selected fuzzy PSO algorithms using common benchmark functions,
- we apply the Takagi–Sugeno fuzzy system [23] which is more computationally efficient than the Mamdani system [16].

In our research, we consider six different versions of PSO, including two non-fuzzy, and four fuzzy algorithms. The evaluation is based on nonlinear benchmarks in the form of Ackley, Griewank, Rastrigin and Rosenbrock functions. The calculations are conducted using Matlab software and the "PSO Research Toolbox" by Evers [6].

2 Particle Swarm Optimization

The particle swarm model consists of particles that are randomly initialized in the d-dimensional search space. The particles explore this space during an iterative process and exchange information to find the optimal solution. Each i-th particle is described by its position $\mathbf{x}_i$, velocity $\mathbf{v}_i$, and personal best position $\mathbf{pbest}_i$. Moreover, the particles have access to the best global position $\mathbf{gbest}$ that has been found by any particle in the swarm.

In the basic PSO algorithm [10], the velocity and the position of each particle in the k-th step of iteration are updated according to the equations

$$\mathbf{v}_i^{k+1} = \mathbf{v}_i^k + c_1 \mathbf{r}_1 (\mathbf{pbest}_i^k - \mathbf{x}_i^k) + c_2 \mathbf{r}_2 (\mathbf{gbest}^k - \mathbf{x}_i^k) \tag{1}$$

$$\mathbf{x}_i^{k+1} = \mathbf{x}_i^k + \mathbf{v}_i^{k+1} \tag{2}$$

where $\mathbf{r}_1$, $\mathbf{r}_2$ are vectors with uniformly distributed random numbers in the interval [0, 1], and c_1, c_2 are positive constants equal to 2.

The velocity of particles is determined by three components. The first component is the inertia that models the particle's tendency to continue moving in the same direction. The second component is cognitive and attracts particles towards their personal best positions. The last component is a social component that moves particles towards the best position found earlier by any particle. Selection of the best global position and the best position for i-th particle is based on the objective function (denoted later by $f(\cdot)$).

2.1 PSO1: Clerc, Kennedy Algorithm [4]

Many approaches have been developed to improve the performance of the basic PSO algorithm. One of them is to use the constriction factor χ that was proposed by Clerc and Kennedy [4]. The application of this factor controls the velocity magnitude.

The velocity equation has the form

$$\mathbf{v}_i^{k+1} = \chi[\mathbf{v}_i^k + c_1\mathbf{r}_1(\mathbf{pbest}_i^k - \mathbf{x}_i^k) + c_2\mathbf{r}_2(\mathbf{gbest}^k - \mathbf{x}_i^k)] \tag{3}$$

where χ is calculated as $\chi = \dfrac{2}{|2-\varphi-\sqrt{\varphi^2-4\varphi}|}$ and $\varphi = c_1 + c_2$, $\varphi > 4$. In this paper, the following typical values are used: $c_1 = c_2 = 2.05$, $\varphi = 4.1$ and $\chi = 0.7298$.

2.2 PSO2: Eberhart, Shi Algorithm [5]

Another way to improve the performance of PSO is to use the inertia weight ω. This parameter is significant because it balances the global exploration and local exploitation abilities of the swarm. Exploration is facilitated when the inertia weight is high, but convergence is slower. On the other hand, when the inertia weight is low then convergence is faster, but it sometimes leads to local solutions. Hence, linearly decreasing inertia weight is proposed in [5].

The velocity equation has the form of

$$\mathbf{v}_i^{k+1} = \omega^k\mathbf{v}_i^k + c_1\mathbf{r}_1(\mathbf{pbest}_i^k - \mathbf{x}_i^k) + c_2\mathbf{r}_2(\mathbf{gbest}^k - \mathbf{x}_i^k) \tag{4}$$

The inertia weight ω is calculated from the formula

$$\omega^k = \omega_{max} - \frac{\omega_{max} - \omega_{min}}{iter_{max}} \cdot k \tag{5}$$

where ω_{max} is the initial weight, ω_{min} is the final weight and $iter_{max}$ is the maximum number of iterations. The limits for ω are set to $\omega_{max} = 0.9$ and $\omega_{min} = 0.4$.

3   Takagi–Sugeno System

Consider the Takagi–Sugeno (T–S) fuzzy system [23] with two inputs y_1, y_2 and one output u. For the input y_1 we define m triangular fuzzy sets A_i (Fig. 1), for which the vertices are placed in points p_i, where $i = 1, \ldots, m$. Similarly, for the input y_2, we define n fuzzy sets B_j with vertices in points q_j, where $j = 1, \ldots, n$. The membership grades $A_i(y_1)$ and $B_j(y_2)$ are calculated from the formulas

$$A_i(y_1) = \text{triang}(y_1; p_{i-1}, p_i, p_{i+1})$$
$$= \max\left(0, \min\left(\frac{y_1 - p_{i-1}}{p_i - p_{i-1}}, \frac{p_{i+1} - y_1}{p_{i+1} - p_i}\right)\right) \tag{6}$$

and

$$B_j(y_2) = \text{triang}(y_2; q_{j-1}, q_j, q_{j+1})$$
$$= \max\left(0, \min\left(\frac{y_2 - q_{j-1}}{q_j - q_{j-1}}, \frac{q_{j+1} - y_2}{q_{j+1} - q_j}\right)\right) \tag{7}$$

For the first and the last fuzzy sets we assign the memberships grades equal to unity if $y_1 \le p_1$ or $y_1 \ge p_m$, $y_2 \le q_1$ or $y_2 \ge q_n$. The coordinates p_i and q_j are written in the form of the vectors $\mathbf{p} = [p_i] = [p_1, \ldots, p_m]$ and $\mathbf{q} = [q_j] = [q_1, \ldots, q_n]$, respectively.

The T–S system is described by $m \cdot n$ fuzzy inference rules of the form

$$R_{ij} : \text{ IF } y_1 \in A_i \text{ AND } y_2 \in B_j, \text{ THEN } u = z_{ij} \tag{8}$$

where $z_{ij} \in \mathbb{R}$ is the consequent of the rule R_{ij}. The rules (8) are written in the following table:

$y_1 \backslash y_2$	B_1	B_2	$\cdots$	B_{n-1}	B_n	
A_1	z_{11}	z_{12}	$\cdots$	$z_{1,n-1}$	z_{1n}	
A_2	z_{21}	z_{22}	$\cdots$	$z_{2,n-1}$	z_{2n}	
$\vdots$	$\vdots$	$\vdots$	$\ddots$	$\vdots$	$\vdots$	(9)
A_{m-1}	$z_{m-1,1}$	$z_{m-1,2}$	$\cdots$	$z_{m-1,n-1}$	$z_{m-1,n}$	
A_m	z_{m1}	z_{m2}	$\cdots$	$z_{m,n-1}$	z_{mn}	

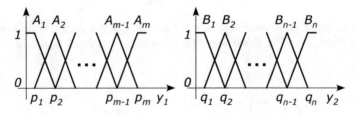

Fig. 1 Fuzzy sets for the inputs y_1 and y_2

The output u is calculated as the weighted average of the consequents z_{ij} and determined by

$$u = TS(y_1, y_2) = \frac{\sum_{i=1}^{m} \sum_{j=1}^{n} w_{ij}(y_1, y_2) z_{ij}}{\sum_{i=1}^{m} \sum_{j=1}^{n} w_{ij}(y_1, y_2)} \tag{10}$$

where $w_{ij}(y_1, y_2) = A_i(y_1) \cdot B_j(y_2)$ denotes the degree of fulfillment of the rule R_{ij}.

4 Fuzzy PSO

4.1 FPSO1: Algorithm Based on the Work by Shi, Eberhart [21]

The better PSO performance can be achieved using the nonlinearly changing inertia weight that balances global and local search abilities. Because it is difficult to design a mathematical model to adapt the inertia weight, therefore this problem may be solved using a linguistic description of the search process. For example, we can use a fuzzy inference system for tuning the inertia weight dynamically [21].

In the FPSO1 algorithm, the inertia weight is described by the formula

$$\omega^{k+1} = \omega^k + \Delta\omega \tag{11}$$

where the change of inertia weight is determined by the T–S fuzzy system (10):

$$\Delta\omega = TS(nf^k, \omega^k) \tag{12}$$

The input nf^k is the normalized objective function value described by

$$nf^k = \frac{fg^k - f_{min}}{f_{max} - f_{min}} \tag{13}$$

where $fg^k = f(\mathbf{gbest}^k)$, f_{min} is the optimal solution (for the benchmark functions applied in this paper, it is equal to 0), f_{max} is the worst solution (in our paper $f_{max} = f(\mathbf{gbest}^0)$). The fuzzy sets for the inputs nf^k and ω^k have vertices in points $\mathbf{p} = [0, 0.5, 1]$, $\mathbf{q} = [0.4, 0.7, 1]$, respectively, and the fuzzy rules are of the form

$$\begin{array}{c|ccc} nf^k \backslash \omega^k & B_1 & B_2 & B_3 \\ \hline A_1 & Z & N & N \\ A_2 & P & Z & N \\ A_3 & P & Z & N \end{array} \tag{14}$$

where $N = -0.1$, $Z = 0$ and $P = 0.1$. For example, the rule R_{11} means that *if nf^k is about 0 and ω^k is about 0.4, then* $\Delta\omega$ is 0.

4.2 FPSO2: Algorithm Based on the Work by Alfi, Fateh [3]

The improvement of the FPSO1 algorithm was proposed by Alfi and Fateh [3]. In their method, the inertia weight is calculated for each particle according to its current state. This is justified because each particle in the swarm is in a different place in a complex environment and may have a different balance between global and local search abilities.

In the FPSO2 algorithm, the T–S system (10) is used to obtain the change of inertia weight for each particle:

$$\Delta\omega_i = TS(nf_i^k, \omega_i^k) \tag{15}$$

where

$$nf_i^k = \frac{fp_i^k - f_{min}}{fp_i^0 - f_{min}} \tag{16}$$

and $fp_i^k = f(\mathbf{pbest}_i^k)$. The vertices of fuzzy sets for nf_i^k and ω_i^k are chosen as $\mathbf{p} = [0, 0.5, 1]$, $\mathbf{q} = [0.4, 0.6, 0.8]$ respectively, and the fuzzy rules are of the form

$$
\begin{array}{c|ccc}
nf_i^k \backslash \omega_i^k & B_1 & B_2 & B_3 \\
\hline
A_1 & P & N & N \\
A_2 & P & Z & N \\
A_3 & P & Z & N \\
\end{array}
\tag{17}
$$

where $N = -0.1$, $Z = 0$ and $P = 0.1$. For example, the rule R_{13} means that *if nf_i^k is about 0 and ω_i^k is about 0.8, then $\Delta\omega_i$ is −0.1.*

4.3 FPSO3: Algorithm Based on the Work by Niknam [19]

In the FPSO3 algorithm, a fuzzy system proposed by Niknam [19] is used to change not only ω, but also the coefficients c_1 and c_2. These coefficients determine the influence of the personal best position $\mathbf{pbest}_i$ and the global best position $\mathbf{gbest}$ on the particle velocity. For example, if c_1 is larger than c_2, then the particle has the tendency to move to the personal best position, rather than to the global best position found by the swarm.

In the FPSO3 algorithm, three T–S systems (10) are used to determine the parameters ω, c_1 and c_2:

$$\omega = TS(nf^k, nu^k) \tag{18}$$

$$c_1 = TS(nf^k, nu^k) \tag{19}$$

$$c_2 = TS(nf^k, nu^k) \tag{20}$$

The input nf^k is defined in (13) and nu^k is the normalized number of iterations without change of the best global position:

$$nu^k = \frac{u^k - u_{min}}{u_{max} - u_{min}} \tag{21}$$

where u^k is the number of iterations without change of the best global position, $u_{min} = 0$ and $u_{max} = iter_{max}$. The fuzzy sets are defined by the vectors $\mathbf{p} = [0.2, 0.4, 0.6, 0.8]$, $\mathbf{q} = [0.2, 0.4, 0.6, 0.8]$. The fuzzy rules for the inertia weight ω have the form of

$$
\begin{array}{c|cccc}
nf^k \backslash nu^k & B_1 & B_2 & B_3 & B_4 \\
\hline
A_1 & PS & PM & PB & PB \\
A_2 & PM & PM & PB & PR \\
A_3 & PB & PB & PB & PR \\
A_4 & PB & PB & PR & PR \\
\end{array}
\tag{22}
$$

In table (22) we have $PS = 0.4$, $PM = 0.6$, $PB = 0.8$ and $PR = 1$. The fuzzy rules for the parameter c_1 are defined as

$$
\begin{array}{c|cccc}
nf^k \backslash nu^k & B_1 & B_2 & B_3 & B_4 \\
\hline
A_1 & PR & PB & PB & PB \\
A_2 & PB & PM & PM & PS \\
A_3 & PB & PM & PS & PS \\
A_4 & PM & PM & PS & PS \\
\end{array}
\tag{23}
$$

and for the parameter c_2 they are defined as

$$
\begin{array}{c|cccc}
nf^k \backslash nu^k & B_1 & B_2 & B_3 & B_4 \\
\hline
A_1 & PR & PB & PM & PM \\
A_2 & PB & PM & PS & PS \\
A_3 & PM & PM & PS & PS \\
A_4 & PM & PS & PS & PS \\
\end{array}
\tag{24}
$$

In tables (23) and (24) we have $PS = 1.2$, $PM = 1.4$, $PB = 1.6$ and $PR = 1.8$.

4.4 MFPSO: Authors' Proposition

The modified fuzzy PSO (MFPSO) algorithm proposed by the authors [11] combines the previously described concepts developed by Alfi, Fateh [3] and Niknam [19]. In this algorithm, each of particles has its own coefficients ω, c_1 and c_2 changing according to the linguistic description represented by the fuzzy rules. In this way, each of the particles may be treated individually. For example, if a particle has found the new local best position $\mathbf{pbest}_i$, then the inertia weight ω should be decreased and the coefficients c_1 and c_2 should be increased. On the other hand, if **pbest**$_i$ has not changed for a long time, ω should be increased and c_1, c_2 should be decreased to improve the ability of exploration.

In the MFPSO algorithm, the authors propose the parameters ω, c_1 and c_2 for each particle to be determined using three T–S systems (10):

$$\omega_i = TS(nf_i^k, nu_i^k) \tag{25}$$

$$(c_1)_i = TS(nf_i^k, nu_i^k) \tag{26}$$

$$(c_2)_i = TS(nf_i^k, nu_i^k) \tag{27}$$

where nf_i^k is defined in (16), nu_i^k has the form of

$$nu_i^k = \frac{u_i^k - u_{min}}{u_{max} - u_{min}} \tag{28}$$

and u_i^k is the number of iterations without change to the best personal position for the i-th particle. It should be noted that in Eq. (21), nu^k is calculated on the basis of the global best position **gbest**, whereas in Eq. (28), nu_i^k is calculated on the basis of the personal best position **pbest**$_i$. The vertices of fuzzy sets for nf_i^k and nu_i^k are defined as $\mathbf{p} = [0.2, 0.45, 0.65, 0.9]$, $\mathbf{q} = [0.2, 0.45, 0.65, 0.9]$.

The fuzzy rules for the inertia weight ω are the same as in (22). The fuzzy rules for the parameters c_1 and c_2 are given in tables (23) and (24), where $PS = 1.5$, $PM = 1.7$, $PB = 1.9$ and $PR = 2.1$. For example, the rule R_{11} has the form

$$\begin{aligned} R_{11} : \text{IF } nf_i^k &\in A_1 \text{ AND } nu_i^k \in B_1, \\ \text{THEN } \omega &= PS \text{ AND } c_1 = PR \text{ AND } c_2 = PR \end{aligned} \tag{29}$$

and the rule R_{23} has the form

$$\begin{aligned} R_{23} : \text{IF } nf_i^k &\in A_2 \text{ AND } nu_i^k \in B_3, \\ \text{THEN } \omega &= PB \text{ AND } c_1 = PM \text{ AND } c_2 = PS \end{aligned} \tag{30}$$

Other rules can be interpreted similarly.

5 Results and Discussion

In order to evaluate the algorithms, four common nonlinear benchmark functions [13, 21] were used:

- Ackley function

$$f_1(\mathbf{x}) = -20 \exp\left(-0.2 \sqrt{\frac{1}{d} \sum_{i=1}^{d} x_i^2}\right) - \exp\left(\frac{1}{d} \sum_{i=1}^{d} \cos(2\pi x_i)\right) + 20 + e \quad (31)$$

- Griewank function

$$f_2(\mathbf{x}) = \frac{1}{4000} \sum_{i=1}^{d} x_i^2 - \prod_{i=1}^{d} \cos\left(\frac{x_i}{\sqrt{i}}\right) + 1 \quad (32)$$

- Rastrigin function

$$f_3(\mathbf{x}) = \sum_{i=1}^{d} (x_i^2 - 10\cos(2\pi x_i) + 10) \quad (33)$$

- Rosenbrock function

$$f_4(\mathbf{x}) = \sum_{i=1}^{d-1} \left(100\,(x_{i+1} - x_i^2)^2 + (x_i - 1)^2\right) \quad (34)$$

where d is the dimension. For these functions, the asymmetric initialization method, as in the paper [21], was used. The velocity of particles was clamped to v_{max}, however, the position of the particles was not limited. The initialization ranges and v_{max} for the test functions are listed in Table 1. In our experiments, three dimension sizes were chosen: $d = 10$, $d = 30$ and $d = 100$. The number of iterations was set to 1000, 2000 and 5000 corresponding to the dimensions 10, 30 and 100. The number of particles was equal to 30 and the number of trials was equal to 30 in all experiments. The calculations were conducted using Matlab software and the "PSO Research Toolbox" by Evers [6]. The average time of execution for one trial of the MFPSO algorithm with 30 particles and 1000 iterations on a mobile workstation equipped with Intel(R) Core(TM) i7-2820QM was about 5 s.

The objective function values for **gbest** for trials in which the algorithms achieved the value closest to the mean are shown in Figs. 2 and 3. The results for the considered

Table 1 Parameters of benchmark functions	Function	Initialization ranges	v_{max}
	Ackley	$(15, 30)^d$	30
	Griewank	$(300, 600)^d$	600
	Rastrigin	$(2.56, 5.12)^d$	5.12
	Rosenbrock	$(15, 30)^d$	30

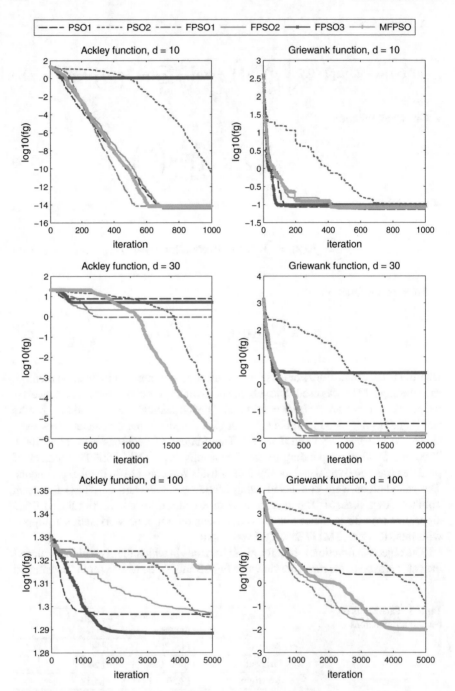

Fig. 2 Objective function values for **gbest** for trials in which the algorithms achieved the value closest to the mean

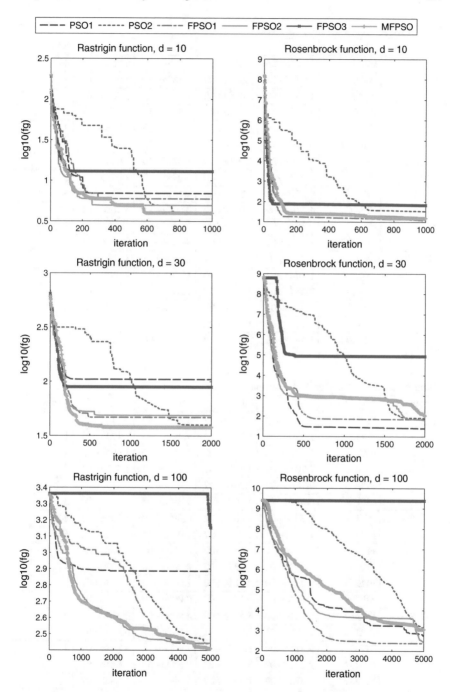

Fig. 3 Objective function values for **gbest** for trials in which the algorithms achieved the value closest to the mean

Table 2 Results for the Ackley function

d, iter	Algorithm	Mean	Sd	Min	Max	Rating
10,1000	PSO1	2.223e−05	1.218e−04	3.553e−15	6.669e−04	4
	PSO2	6.685e−01	3.662e+00	8.882e−14	2.006e+01	3
	FPSO1	1.332e+00	5.068e+00	3.553e−15	2.006e+01	1
	FPSO2	3.790e−15	9.013e−16	3.553e−15	7.105e−15	6
	FPSO3	9.000e−01	3.662e+00	3.553e−15	2.006e+01	2
	MFPSO	5.566e−15	2.412e−15	3.553e−15	1.421e−14	5
30,2000	PSO1	8.218e+00	7.791e+00	1.421e−14	1.980e+01	2
	PSO2	6.909e−01	3.784e+00	6.994e−07	2.073e+01	6
	FPSO1	9.953e−01	3.781e+00	1.421e−14	2.079e+01	5
	FPSO2	4.146e+00	8.032e+00	2.807e−13	2.003e+01	3
	FPSO3	8.351e+00	7.650e+00	1.344e+00	1.998e+01	1
	MFPSO	4.104e+00	8.348e+00	8.846e−11	2.087e+01	4
100,5000	PSO1	1.980e+01	4.430e−02	1.970e+01	1.992e+01	4
	PSO2	1.696e+01	6.888e+00	1.708e+00	2.067e+01	6
	FPSO1	2.038e+01	2.155e+00	1.097e+01	2.112e+01	1
	FPSO2	1.983e+01	1.933e−01	1.899e+01	2.017e+01	3
	FPSO3	1.934e+01	2.292e+00	7.226e+00	1.988e+01	5
	MFPSO	1.993e+01	2.799e+00	9.303e+00	2.114e+01	2

Table 3 Results for the Griewank function

d, iter	Algorithm	Mean	Sd	Min	Max	Rating
10,1000	PSO1	7.258e−02	3.584e−02	3.197e−02	2.115e−01	6
	PSO2	1.050e−01	5.726e−02	7.396e−03	2.172e−01	1
	FPSO1	8.642e−02	3.961e−02	1.969e−02	1.796e−01	4
	FPSO2	7.701e−02	3.531e−02	3.201e−02	1.847e−01	5
	FPSO3	9.796e−02	5.344e−02	1.970e−02	2.511e−01	2
	MFPSO	8.690e−02	4.569e−02	9.857e−03	2.017e−01	3
30,2000	PSO1	2.701e−02	4.005e−02	0.000e+00	1.858e−01	2
	PSO2	1.303e−02	1.775e−02	2.092e−11	9.064e−02	6
	FPSO1	1.375e−02	1.688e−02	0.000e+00	5.888e−02	5
	FPSO2	1.492e−02	2.037e−02	0.000e+00	9.562e−02	4
	FPSO3	3.036e+00	2.411e+00	1.049e+00	1.128e+01	1
	MFPSO	1.661e−02	2.176e−02	0.000e+00	8.558e−02	3
100,5000	PSO1	2.841e+00	6.539e+00	4.998e−03	2.589e+01	2
	PSO2	1.186e−01	1.255e−01	2.093e−02	6.304e−01	3
	FPSO1	9.915e−02	1.467e−01	1.056e−11	6.202e−01	4
	FPSO2	2.237e−02	3.622e−02	9.593e−09	1.645e−01	5
	FPSO3	4.572e+02	1.202e+02	1.990e+02	6.533e+02	1
	MFPSO	9.624e−03	1.698e−02	1.361e−05	6.498e−02	6

Table 4 Results for the Rastrigin function

d, iter	Algorithm	Mean	Sd	Min	Max	Rating
10,1000	PSO1	7.097e+00	3.969e+00	1.990e+00	1.890e+01	2
	PSO2	3.715e+00	1.865e+00	0.000e+00	7.960e+00	6
	FPSO1	5.804e+00	2.575e+00	2.985e+00	1.293e+01	3
	FPSO2	4.743e+00	3.394e+00	0.000e+00	1.791e+01	4
	FPSO3	1.270e+01	5.817e+00	9.950e−01	2.388e+01	1
	MFPSO	4.053e+00	2.627e+00	4.421e−03	8.955e+00	5
30,2000	PSO1	1.053e+02	2.743e+01	4.676e+01	1.512e+02	1
	PSO2	3.819e+01	9.564e+00	2.389e+01	6.766e+01	6
	FPSO1	4.580e+01	8.148e+00	3.084e+01	6.368e+01	4
	FPSO2	4.852e+01	1.391e+01	2.388e+01	8.457e+01	3
	FPSO3	9.155e+01	2.314e+01	5.330e+01	1.353e+02	2
	MFPSO	4.024e+01	1.019e+01	1.293e+01	5.771e+01	5
100,5000	PSO1	7.682e+02	1.328e+02	4.636e+02	1.094e+03	2
	PSO2	2.806e+02	4.189e+01	2.062e+02	3.612e+02	3
	FPSO1	2.781e+02	3.371e+01	2.040e+02	3.512e+02	4
	FPSO2	2.776e+02	4.333e+01	1.991e+02	3.691e+02	5
	FPSO3	1.197e+03	7.490e+02	5.457e+02	2.344e+03	1
	MFPSO	2.627e+02	4.685e+01	1.733e+02	3.384e+02	6

Table 5 Results for the Rosenbrock function

d, iter	Algorithm	Mean	Sd	Min	Max	Rating
10,1000	PSO1	2.155e+01	3.702e+01	1.381e−02	1.261e+02	4
	PSO2	3.602e+01	1.308e+02	6.977e−01	7.244e+02	2
	FPSO1	1.761e+01	3.553e+01	2.459e−03	1.371e+02	5
	FPSO2	2.529e+01	5.990e+01	7.227e−02	2.577e+02	3
	FPSO3	7.058e+01	2.101e+02	2.104e+00	1.152e+03	1
	MFPSO	1.742e+01	5.445e+01	8.144e−03	2.854e+02	6
30,2000	PSO1	3.793e+01	5.813e+01	6.057e−02	2.642e+02	6
	PSO2	8.484e+01	7.490e+01	5.490e+00	3.359e+02	3
	FPSO1	6.247e+01	7.706e+01	3.930e−01	3.082e+02	4
	FPSO2	5.779e+01	4.336e+01	1.420e+00	1.683e+02	5
	FPSO3	8.847e+04	1.825e+05	2.877e+02	8.705e+05	1
	MFPSO	9.524e+01	1.533e+02	5.333e+00	8.481e+02	2
100,5000	PSO1	4.937e+02	4.707e+02	1.488e+02	2.073e+03	4
	PSO2	1.360e+03	1.526e+03	4.846e+02	8.401e+03	2
	FPSO1	2.390e+02	9.148e+01	9.722e+01	6.008e+02	6
	FPSO2	2.653e+02	9.099e+01	1.230e+02	5.349e+02	5
	FPSO3	2.416e+09	4.259e+08	2.170e+08	2.635e+09	1
	MFPSO	1.177e+03	1.916e+03	3.616e+02	1.078e+04	3

Table 6 Ratings of the PSO algorithms

Algorithm	d = 10	d = 30	d = 100	$\sum$
PSO1	16	11	12	39
PSO2	12	21	14	47
FPSO1	13	18	15	46
FPSO2	18	15	18	51
FPSO3	6	5	8	19
MFPSO	19	14	17	50

benchmark functions are presented in Tables 2, 3, 4 and 5. These tables contain the basic statistics for the final value of the objective function and the ratings of the algorithms. These ratings were assigned in such a way that the best algorithm received six points and the worst received one point. The ratings of the algorithms are summarized in Table 6. For the dimension $d = 10$, the highest rating has the algorithm MFPSO proposed by the authors. For $d = 30$ and $d = 100$ the highest ratings have the algorithms PSO2 and FPSO2, respectively. The MFPSO achieved the fourth result in case of $d = 30$ and the second result in case of $d = 100$. Analyzing the sum of ratings it can be seen that the best is the FPSO2 algorithm and the second is the MFPSO algorithm.

6 Conclusion

In this paper, a comparison of selected fuzzy particle swarm optimization algorithms was presented. A modified fuzzy PSO algorithm was proposed, in which each of the particles has its own inertia weight and the coefficients of the cognitive and social components. The Takagi–Sugeno fuzzy system was used instead of the Mamdani fuzzy system because it has a shorter processing time. Six different versions of PSO were considered, including two non-fuzzy and four fuzzy algorithms. The evaluation was based on nonlinear benchmarks in the form of Ackley, Griewank, Rastrigin and Rosenbrock functions. The calculations were conducted using Matlab software and the "PSO Research Toolbox" [6].

Further work will focus on improving the proposed algorithm, its application in tracking objects in images, the analysis of athletes' technique [12], and building models to support the training process in sport [25].

References

1. Abdelbar, A.M., Abdelshahid, S., Wunsch, D.C.: Fuzzy PSO: a generalization of particle swarm optimization. In: Proceedings. IEEE International Joint Conference on Neural Networks, vol. 2, pp. 1086–1091 (2005). doi:10.1109/IJCNN.2005.1556004

2. Adamczyk, M.: Parallel feature selection algorithm based on rough sets and particle swarm optimization. In: 2014 Federated Conference on Computer Science and Information Systems (FedCSIS), pp. 43–50 (2014). doi:10.15439/2014F389

3. Alfi, A., Fateh, M.M.: Intelligent identification and control using improved fuzzy particle swarm optimization. Expert Syst. Appl. **38**(10), 12312–12317 (2011). doi:10.1016/j.eswa.2011.04. 009

4. Clerc, M., Kennedy, J.: The particle swarm - explosion, stability, and convergence in a multidimensional complex space. IEEE Trans. Evol. Comput. **6**(1), 58–73 (2002). doi:10.1109/4235. 985692

5. Eberhart, R.C., Shi, Y.: Comparing inertia weights and constriction factors in particle swarm optimization. In: Evolutionary Computation, 2000. Proceedings of the 2000 Congress on, vol. 1, pp. 84–88 (2000). doi:10.1109/CEC.2000.870279

6. Evers, G.: PSO Research Toolbox (Version 20110515), M.S. thesis code (2016). http://www. georgeevers.org/pso_research_toolbox.htm

7. Izakian, H., Abraham, A., Snášel, V.: Fuzzy clustering using hybrid fuzzy c-means and fuzzy particle swarm optimization. In: 2009 World Congress on Nature Biologically Inspired Computing (NaBIC), pp. 1690–1694 (2009). doi:10.1109/NABIC.2009.5393618

8. Juang, Y.T., Tung, S.L., Chiu, H.C.: Adaptive fuzzy particle swarm optimization for global optimization of multimodal functions. Inf. Sci. **181**(20), 4539–4549 (2011). Special Issue on Interpretable Fuzzy Systems. doi:10.1016/j.ins.2010.11.025

9. Karami, A., Guerrero-Zapata, M.: A fuzzy anomaly detection system based on hybrid PSO-Kmeans algorithm in content-centric networks. Neurocomputing **149, Part C**, 1253–1269 (2015). doi:10.1016/j.neucom.2014.08.070

10. Kennedy, J., Eberhart, R.: Particle swarm optimization. In: Proceedings of IEEE International Conference on Neural Networks, vol. 4, pp. 1942–1948. IEEE Press, Piscataway, NJ (1995). doi:10.1109/ICNN.1995.488968

11. Krzeszowski, T., Wiktorowicz, K.: Evaluation of selected fuzzy particle swarm optimization algorithms. In: 2016 Federated Conference on Computer Science and Information Systems (FedCSIS), pp. 571–575 (2016). doi:10.15439/2016F206

12. Krzeszowski, T., Przednowek, K., Wiktorowicz, K., Iskra, J.: Estimation of hurdle clearance parameters using a monocular human motion tracking method. Comput. Methods Biomech. Biomed. Eng. **19**(12), 1319–1329 (2016). doi:10.1080/10255842.2016.1139092

13. Liang, J.J., Suganthan, P.N., Deb, K.: Novel composition test functions for numerical global optimization. In: Proceedings 2005 IEEE Swarm Intelligence Symposium, 2005. SIS 2005., pp. 68–75 (2005). doi:10.1109/SIS.2005.1501604

14. Ling, S.H., Nguyen, H.T., Leung, F.H.F., Chan, K.Y., Jiang, F.: Intelligent fuzzy particle swarm optimization with cross-mutated operation. In: 2012 IEEE Congress on Evolutionary Computation, pp. 1–8 (2012). doi:10.1109/CEC.2012.6252934

15. Liu, H., Abraham, A., Zhang, W.: A fuzzy adaptive turbulent particle swarm optimisation. Int. J. Innov. Comput. Appl. **1**(1), 39–47 (2007). doi:10.1504/IJICA.2007.013400

16. Mamdani, E., Assilian, S.: An experiment in linguistic synthesis with a fuzzy logic controller. Int. J. Man Mach. Stud. **7**(1), 1–13 (1975). doi:10.1016/S0020-7373(75)80002-2

17. Mohiuddin, M.A., Khan, S.A., Engelbrecht, A.P.: Fuzzy particle swarm optimization algorithms for the open shortest path first weight setting problem. Appl. Intell. **45**(3), 598–621 (2016). doi:10.1007/s10489-016-0776-0

18. Nesamalar, J.J.D., Venkatesh, P., Raja, S.C.: Managing multi-line power congestion by using Hybrid Nelder-Mead - Fuzzy Adaptive Particle Swarm Optimization (HNM-FAPSO). Appl. Soft Comput. **43**, 222–234 (2016). doi:10.1016/j.asoc.2016.02.013

19. Niknam, T.: A new fuzzy adaptive hybrid particle swarm optimization algorithm for non-linear, non-smooth and non-convex economic dispatch problem. Appl. Energy **87**(1), 327–339 (2010). doi:10.1016/j.apenergy.2009.05.016

20. Saini, S., Zakaria, N., Rambli, D.R.A., Sulaiman, S.: Markerless human motion tracking using hierarchical multi-swarm cooperative particle swarm optimization. PLoS ONE **10**(5) (2015). doi:10.1371/journal.pone.0127833

21. Shi, Y., Eberhart, R.C.: Fuzzy adaptive particle swarm optimization. Proc. Congr. Evol. Comput. **1**, 101–106 (2001). doi:10.1109/CEC.2001.934377
22. Srinivasan, D., Loo, W.H., Cheu, R.L.: Traffic incident detection using particle swarm optimization. In: Proceedings of the IEEE Swarm Intelligence Symposium. SIS '03, pp. 144–151 (2003). doi:10.1109/SIS.2003.1202260
23. Takagi, T., Sugeno, M.: Fuzzy identification of systems and its applications to modeling and control. IEEE Trans. Syst. Man Cybern. **SMC-15**(1), 116–132 (1985). doi:10.1109/TSMC.1985.6313399
24. Tian, D.P., Li, N.Q.: Fuzzy particle swarm optimization algorithm. In: 2009 International Joint Conference on Artificial Intelligence, pp. 263–267 (2009). doi:10.1109/JCAI.2009.50
25. Wiktorowicz, K., Przednowek, K., Lassota, L., Krzeszowski, T.: Predictive modeling in race walking. Comput. Intell. Neurosci. **2015**, 9 (2015). doi:10.1155/2015/735060. Article ID 735060

On the Exact Solution of the Distance Geometry with Interval Distances in Dimension 1

Antonio Mucherino

Abstract Distance Geometry consists in embedding a simple weighted undirected graph in a given space so that the distances between embedded vertices correspond to the edge weights. Weights can be either exact real values, or real-valued intervals. In this work, the focus is on problems where the embedding space is the Euclidean 1-dimensional space, and the general situation where distances can be represented by intervals is taken into consideration. A previously proposed branch-and-prune algorithm is adapted to the 1-dimensional case, and the proposed variant turns out to be deterministic even in presence of interval distances. Backtracking pruning is introduced in the algorithm for guaranteeing that all vertex positions in a given solution are actually feasible. The proposed algorithm is tested on a set of artificially generated instances in dimension 1.

1 Introduction

Given a simple weighted undirected graph $G = (V, E, d)$ and a dimension $K > 0$, the Distance Geometry Problem (DGP) asks whether an embedding $x : V \longrightarrow \mathbb{R}^K$ of G exists in the K-dimensional Euclidean space $\mathbb{R}^K$ so that the following distance constraints are satisfied:

$$\forall \{u, v\} \in E, \quad ||x_u - x_v|| \in d_{uv}.$$

Notice that the symbol "∈" can stand either for an equality, when the distance d_{uv} is exact, or rather for two inequalities, in which case d_{uv} is actually represented by a real-valued interval $[d_{uv}^L, d_{uv}^U]$. The subset of E corresponding to the exact distances in E will be referred in the following with the symbol E'. The focus of this short paper is on DGPs arising in dimension $K = 1$ and for which all distances are represented by intervals (the set E' may therefore be empty). The DGP is NP-complete in dimension $K = 1$, and strongly NP-hard in any other dimensions [21].

A. Mucherino (✉)
IRISA, University of Rennes 1, Rennes, France
e-mail: antonio.mucherino@irisa.fr

© Springer International Publishing AG 2018   123
S. Fidanova (ed.), *Recent Advances in Computational Optimization*,
Studies in Computational Intelligence 717, DOI 10.1007/978-3-319-59861-1_8

Several real-life problems can be formulated as a DGP [15, 18, 19]. Classical problems include the one of finding the location of sensors in a given network by exploiting point-to-point distance approximations [5, 7, 22], and the one of identifying the conformation of molecules from interatomic distances obtained through experimental techniques [1, 6, 12]. While the embedding space has dimension 3 in the latter, the dimension can be either 2 or 3 in the former application.

In dimension 1, the clock synchronization problem can be formulated as a DGP [8, 23]. The problem consists in computing the internal clock time for machines in a given network by exploiting their own offset with respect to a predefined clock, which is used as a reference. When all offsets are precisely provided, the identification of solutions can be performed by employing a discrete approach to the DGP, as far as there exists a vertex order on V such that every clock has known offset with a preceding one (see Sect. 2). However, when the offsets are represented by intervals, the DGP search space cannot be discretized by employing existing approaches.

This short paper presents a variant on a branch-and-prune (BP) algorithm. The BP algorithm was initially conceived for discretizable instances of the DGP [14]. The proposed variant, named BP1, is particularly designed for problems arising in dimension 1, where a search space having the structure of a tree can be defined even in presence of interval distances. In this search space, the tree nodes are not vertex positions, but rather an interval of feasible positions for a given vertex can be associated to each tree node. A new feature of BP1 is the *backtracking pruning*, which allows to remove infeasible parts of vertex position intervals when backtracking during the tree search. This way, all generated intervals for vertex positions are feasible with respect to some positions belonging to intervals related to other vertices.

This paper is organized as follows. In Sect. 2, a brief description of the assumptions allowing for the discretization of DGPs, in any dimension, is given, together with a sketch of the general BP algorithm. Section 3 presents the proposed variant of the BP algorithm in dimension 1, while some experiments on artificially generated instances are presented in Sect. 4. Finally, Sect. 5 concludes the paper.

2   Discretizable DGPs

Let $G = (V, E, d)$ be a simple weighted undirected graph representing an instance on the DGP in dimension $K > 0$. It is supposed that a vertex ordering is associated to the vertex set V, so that a unique rank is associated to each vertex (in the following, the vertex and its rank will be denoted with the same symbol). The Discretizable DGP (DDGP) is a class of instances of the DGP that satisfy the following assumptions [11, 12, 17]:

(a) $G[\{1, 2, \ldots, K\}]$ is a clique;
(b) $\forall v \in \{K + 1, \ldots, |V|\}$, there exist K vertices $u_1, u_2, \ldots, u_K \in V$ such that

1. $u_1 < v, u_2 < v, \ldots, u_K < v$;
2. $\{\{u_1, v\}, \{u_2, v\}, \ldots, \{u_{K-1}, v\}\} \subset E'$ and $\{u_K, v\} \in E$;

for which

$$\mathcal{V}_S(u_1, u_2, \ldots, u_K) > 0 \qquad (\text{if } K > 1),$$

where $G[\cdot]$ is the subgraph induced by a subset of vertices of V, and $\mathcal{V}_S(\cdot)$ is the volume of the simplex generated by an embedding of the vertices $u_1, u_2, \ldots, u_K$. Notice that volume-invariant embeddings for such vertices can be identified, before the solution of the instance, as far as they form a K-clique in G; if not, this verification cannot be performed in advance. However, the volume $\mathcal{V}_S$ can be zero with probability 0, and it is therefore common use to neglect this assumption when dealing with real-life instances [12]. Also notice that, for $K = 1$, the simplex reduces to a singleton.

Vertex orders satisfying assumptions (a) and (b) are also named *discretization orders*. Assumption (a) ensures the existence of an initial clique, which can be exploited for fixing the coordinate system for the solutions, in order to avoid to regenerating solutions that can be obtained from others by applying rotations and translations. Assumption (b) allows for reducing the search space for the DGP instance to a discrete domain having the structure of a tree, where the positions of vertices are organized layer by layer. Since the K *reference vertices* $u_1, u_2, \ldots, u_K$ for the current vertex v precede v in the vertex ordering, it is supposed that a position is already available for them, when positions for v are searched. By exploiting the corresponding *reference distances*, K spheres can be defined, whose intersection gives the set of possible positions for the vertex v. This set of positions has cardinality 2 with probability 1 [17]. If one of the reference distances is represented by an interval, then the intersection of the spheres gives two disjoint arcs with probability 1 [12].

The branch-and-prune (BP) algorithm was initially conceived for solving this class of instances [14]. The basic idea is to recursively explore the search tree that can be obtained by applying assumptions (a) and (b) above. For the current vertex v, the algorithm generates either 2 positions, or two arcs. In the first case, the algorithm subsequently invokes itself for the two computed positions. Otherwise, in the second possible situation, samples positions are extracted from each arc, and the algorithm is invoked as many times as the number of chosen sample positions [12]. This is the *branching phase* of BP; its *pruning phase* consists in verifying, when this is possible, the feasibility of the computed vertex positions, by checking whether any additional distances (that were not used yet in the computations) are also satisfied. Depending on which phase of the BP algorithm they are employed, the reference distances can be classified in *discretization distances* and *pruning distances*.

Algorithm 1 is a sketch of the BP algorithm for DDGP instances in dimension $K > 0$. It is supposed that the initial clique is already embedded, and that the search starts from $v = K + 1$. Apart from the vertex v, which is the current one to be embedded, and the information about the graph G, the algorithm accepts an integer and strictly positive value D, that indicates the number of sample positions to be taken from computed arcs. Naturally, when the arc $\mathcal{C}_v^0$, or $\mathcal{C}_v^1$, reduces to a singleton,

---

Algorithm 1 The BP algorithm.

1: BP(v, G, D)
2: **compute** the arc $\mathcal{C}_v^0$ by sphere intersection;
3: **extract** D different sample positions $x_v^{0,\ell}$ from $\mathcal{C}_v^0$;
4: **for each** $\ell \in \{1, 2, \ldots, D\}$ **do**
5:   **if** ($x_v^{0,\ell}$ is feasible wrt the pruning distances) **then**
6:     **if** ($v = |V|$) **then**
7:       **print** solution;
8:     **else**
9:       BP($v + 1$, G, D);
10:    **end if**
11: **end if**
12: **end for**
13: **compute** the arc $\mathcal{C}_v^1$ by sphere intersection;
14: **extract** D different sample positions $x_v^{1,\ell}$ from $\mathcal{C}_v^1$;
15: **for each** $\ell \in \{1, 2, \ldots, D\}$ **do**
16:   **if** ($x_v^{1,\ell}$ is feasible wrt the pruning distances) **then**
17:     **if** ($v = |V|$) **then**
18:       **print** solution;
19:     **else**
20:       BP($v + 1$, G, D);
21:     **end if**
22:   **end if**
23: **end for**

---

then only one position is extracted. For every new call to BP, the vertex positions, computed on the current branch of the tree, are supposed to be kept into the global memory.

It is important to point out that the BP algorithm is deterministic when all available distances are exact. When the pruning distances are allowed to be represented by a real-valued interval (and the discretization distances are kept exact), it was observed empirically that the uncertainty on the pruning distances implies an increase on the total number of found solutions [16]. However, in this situation, the BP algorithm is still deterministic. It becomes a heuristic when at least one interval distance is contained in the set of discretization distances. Even with large values for the parameter D, the selection of a predefined number of sample positions from an arc unavoidably introduces errors, which are likely to propagate over the layers of the search tree.

With the aim of reducing the magnitude of such introduced errors, two main strategies have been proposed in the scientific literature, both aimed at improving the selection procedure of samples positions from the obtained arcs [2–4, 9, 13]. In both cases, the main idea is to select, at level v of the search tree, only the sample positions that are feasible with respect to *all* available distances, and not only to the ones used for defining the spheres in the intersections. In other words, instead of generating vertex positions that are only feasible with respect to the discretization distances (and then to apply the pruning phase), the idea is to perform the pruning phase *during* the branching phase, by selecting only positions that also satisfy the pruning distances.

The description of these two strategies goes beyond the purposes of this paper, and the reader is mainly referred to [9, 13] for additional information. It is important to remark however that both strategies allow to represent the selected arc parts in algebraic form. Therefore, one may consider replacing sample positions with equations, at each node of the search tree. This would overcome the problem of introducing errors while selecting sample positions. However, the centers of the spheres used in the intersections need to be represented by singletons, and not by equations, which makes the sample selection process strictly necessary.

3   A Branch-and-Prune (BP) Algorithm in Dimension 1

Let $G = (V, E, d)$ be a simple weighted undirected graph representing an instance of the DGP in dimension 1 for which there exists a vertex ordering on V such that, for every vertex $v \in V$, there is at least one vertex u that is a reference vertex for v (i.e., $u < v$ and $\{u, v\} \in E$). This is equivalent to requiring that the DGP instance satisfies the DDGP assumptions in dimension 1 (see Sect. 1).

Since the focus is on general problems consisting of interval distances, it is reasonable to associate, to every vertex v, *intervals* of positions (rather than singletons) for which the distance constraints are satisfied. In the following, the notation $[z_v^L, z_v^U]$ will be employed for referring to a suitable interval for v, where z_v^L is its lower bound, and z_v^U is its upper bound. The minimal and the maximal distance between two distinct position intervals $[z_u^L, z_u^U]$ and $[z_v^L, z_v^U]$ can be defined as follows:

$$d_{min}\left([z_u^L, z_u^U], [z_v^L, z_v^U]\right) = \begin{cases} \max\{z_u^L, z_v^L\} - \min\{z_u^U, z_v^U\} & \text{if } [z_u^L, z_u^U] \cap [z_v^L, z_v^U] = \emptyset \\ 0 & \text{otherwise,} \end{cases}$$

$$d_{max}\left([z_u^L, z_u^U], [z_v^L, z_v^U]\right) = \max\{z_u^U, z_v^U\} - \min\{z_u^L, z_v^L\}.$$

Let $\mathcal{I}$ be the set of all intervals in $\mathbb{R}$. When replacing single positions with position intervals for the vertices, it is reasonable to seek a set of mappings

$$z : v \in V \longrightarrow [z_v^L, z_v^U] \in \mathcal{I}$$

such that all distance constraints are satisfied, i.e.:

$$\forall \{u, v\} \in E \quad d_{uv}^L \leq d_{min}\left([z_u^L, z_u^U], [z_v^L, z_v^U]\right) \leq d_{max}\left([z_u^L, z_u^U], [z_v^L, z_v^U]\right) \leq d_{uv}^U.$$

Each mapping z will be referred to as a "BP1 solution"; a set of embeddings x can be identified from a given BP1 solution.

BP1 is a variant on the classical BP algorithm (see Sect. 2) and has two fundamental features: it performs a search over a tree even in presence of interval distances; it implements an additional pruning phase, during backtracking. Since the embedding

---

Algorithm 2 The BP1 algorithm.

1: BP1(v, G)
2: **let** w be the reference vertex of v with rank closest to v;
3: **let** $I_v = [z_w^L - d_{wv}^U, z_w^U - d_{wv}^L] \cup [z_w^L + d_{wv}^L, z_w^U + d_{wv}^U]$;
4: **for** (all other reference vertices u) **do**
5:    **let** $J_1 = [z_u^L - d_{uv}^U, z_u^U - d_{uv}^L]; J_2 = [z_u^L + d_{uv}^L, z_u^U + d_{uv}^U]$;
6:    **let** $I_v = I_v \cap (J_1 \cup J_2)$;
7: **end for**
8: **for** (all reference vertices u (including the initial one: w)) **do**
9:    **let** $Q = I_u \cap \bigcup \left([z_v^L - d_{uv}^U, z_v^U - d_{uv}^L] \cup [z_v^L + d_{uv}^L, z_v^U + d_{uv}^U]\right)$;
10:   **if** $(Q \neq I_u)$ **then**
11:     **let** $back = u$;
12:     **let** $I_u = Q$;
13:   **end if**
14: **end for**
15: **for** (all intervals in I_v) **do**
16:   **if** $(v = |V|)$ **then**
17:     **print** intervals I_* belonging to the current branch;
18:   **else**
19:     **if** $(back = 0)$ **then**
20:       **call** BP1$(v + 1, G)$;
21:     **else if** $(back = v)$ **then**
22:       **recall** BP1(v, G) with the updated I_v;
23:       **let** $back = 0$;
24:     **end if**
25:   **end if**
26: **end for**

---

dimension is 1, the distance constraints define intervals in the 1-dimensional Euclidean space, and the intersection of such intervals (in case more than one constraint concerns the same vertex) still consists in a set of intervals. Branching can therefore be performed over the intervals of the set obtained by performing such intersections. Extreme situations are the ones where this intersection is empty, or it reduces to only one interval (or even to one singleton). In the former situation, the BP1 algorithm, as the classical BP, prunes the current branch and backtracks the search.

A sketch of the BP1 algorithm is given in Algorithm 2. As the general BP algorithm, BP1 accepts as arguments the current vertex v, in the given vertex ordering, together with the graph G. Differently from BP, it is not necessary to specify the value of D in the BP1 call. The fact that the dimension of the Euclidean space is set to 1 makes the computation of the vertex position intervals easy to perform. Let $[z_w^L, z_w^U]$ be an interval for the feasible positions for the vertex w that precedes v in the vertex ordering. If $\{w, v\} \in E$, then the two possible intervals for v that are compatible with w are:

$$[z_w^L - d_{wv}^U, z_w^U - d_{wv}^L], \qquad [z_w^L + d_{wv}^L, z_w^U + d_{wv}^U]. \tag{1}$$

If the two intervals are not disjoint, then their union can be taken into consideration.

In the BP1 algorithm, the two intervals are stored in the set I_v (notice that I_v is in fact a set of intervals, and not a single one). In the situation where another reference vertex u for v exists, then two other intervals of possible positions for v can be generated, and they can be intersected with the ones already included in I_v. The for loop in lines 4–7 computes all position intervals related to every reference vertex, and updates I_v. This process of updating the set I_v with the additional reference vertices u is equivalent to performing the pruning phase in the BP algorithm. As in the two strategies mentioned in Sect. 2, the branching phase is performed only after the verification of all distances concerning the current vertex v, and is implemented at line 15 in BP1. Instead of branching over positions, BP1 branches over disjoint intervals of positions, and performs therefore a search on a tree where, layer by layer, intervals of feasible positions for a given vertex v are kept. When a leaf node is reached, all position intervals on the current tree branch provide a solution to the problem. It is important to point out that BP1 can enumerate the entire solution set.

The novel pruning phase included in BP1 is performed during backtracking. It allows for refining all position intervals in I_u such that u was used as a reference for the vertex $v > u$. The execution of lines 4–7 ensures that all positions for v in I_v are compatible with all reference vertices. However, when proceeding with the exploration of the current branch, it may turn out that some of the vertex positions included in the interval I_v are actually infeasible with respect to subsequent vertices in the order. The role of the *backtracking pruning* is to remove those infeasible positions from the intervals in I_v.

After the execution of lines 4–7, the initial two intervals used for initializing I_v at line 3 and the various intervals J_1 and J_2 defined at line 5 can be either included in I_v, or only partially included in I_v, or not at all. The last two cases are the most interesting ones. If only a subset of J_1, related to a certain reference vertex u, is included in I_v, this implies that the positions belonging to $J_1 \setminus I_v$ are not feasible with respect to all available distances. Since these positions were computed by using as a reference some positions in I_u, these positions are in fact infeasible. The idea therefore is, by "projecting" the final set I_v on the set I_u, to identify the positions in I_u that are not feasible. When the backtracking pruning refines the set I_u, the branch rooted at I_u is removed and BP1 restarts the search from an updated set of intervals. In Algorithm 2, the variable *back* is supposed to be global, and is used for controlling the restarts after the backtracking pruning phase.

Differently from BP, the BP1 algorithm is deterministic, even when working on instances containing interval data. On the other hand, when all distances are exact, BP1 reduces to BP, because in this case the backtracking pruning is implicit in the branching and pruning mechanism of the original BP. As for BP, BP1 can have a linear best-case complexity, and an exponential worst-case complexity.

The BP1 algorithm ensures that, for every vertex u, every position in the interval $[z_u^L, z_u^U]$ is compatible with at least one position in $[z_v^L, z_v^U]$, when $\{u, v\} \in E$. Algorithm 3 gives a sketch of a procedure for extracting an embedding x from a given BP1 solution z. Notice that the backtracking pruning of BP1 makes sure that the sets J_u cannot become empty when updated at line 8 of Algorithm 3.

Algorithm 3 Extracting an embedding x from a given BP1 solution z.

1: Extract(z, x)
2: **for** all $v \in V$ **do**
3:    set $J_v = [z_v^L, z_v^U]$;
4: **end for**
5: **for** all $v \in V$ **do**
6:    **choose** one position x_v from J_v;
7:    **for** all $u \in V$, such that $\{u, v\} \in E$, **do**
8:      let $J_u = J_u \cap \{[x_v - d_{uv}^U, x_v - d_{uv}^L] \cup [x_v + d_{uv}^L, x_v + d_{uv}^U]\}$;
9:    **end for**
10: **end for**

Algorithm 4 An algorithm for generating DGP instances in dimension 1.

1: Instance_Generation(n, p_t's,ε, G)
2: set $V = \{1, 2, \ldots, n\}$;
3: set $E = \emptyset$;
4: set $m = 0$;
5: *randomly* **select** positions x_v for all vertices v in the real interval $[0, 1]$;
6: **for** every v ranked from 2 to n **do**
7:    *randomly* **select** t such that $p_t > 0$;
8:    let $p_t = p_t - 1$;
9:    let $t = \min(t, v)$;
10:   **for** all t **do**
11:     *randomly* **select** a reference vertex u for v (not selected yet if $t > 1$);
12:     let $E = E \cup \{\{u, v\}\}$;
13:     set σ to a random value in $[0, 1]$;
14:     let $d_{uv}^L = |x_v - x_u| - \sigma\varepsilon$ and $d_{uv}^U = |x_v - x_u| + (1 - \sigma)\varepsilon$;
15:   **end for**
16:   let $m = m + t$;
17: **end for**

4 Some Computational Experiments

This section presents some computational experiments on artificially generated instances of the DGP in dimension 1 such that $E' = \emptyset$ (all distances are represented by intervals). For all instances, it is supposed that a vertex ordering exists so that every vertex v has at least one reference vertex. The BP1 algorithm was implemented in C programming language, and compiled with the gcc compiler, version 4.9.2, under Linux. All experiments have been carried out on computer equipped with a 12M Cache, 2.40 GHz, 5.86 GT/s Intel(R) Xeon(R) CPU, running Linux.

The following procedure was implemented for the generation of the instances. Let n be the size of the instance to be created. Let ε be the error introduced on the true distances, when an interval is generated randomly so that it contains a given distance. Let p_t, with $t \in \{1, 2, 3, 4\}$, be four real values representing four percentages ($0 \leq p_t \leq 1$), where $\sum_t p_t = 1$. Every value p_t indicates the percentage of vertices in the instance that will have t reference vertices (in the implicitly created vertex ordering).

Algorithm 4 sketches the used procedure for generating the instances. Given in input the number n of vertices, the percentages p_t multiplied by 100 and a real positive value ε, the procedure outputs a graph G representing an instance of the DGP in dimension 1. The number of reference vertices per vertex, as well as the ranks of these reference vertices, are randomly chosen. The distances that can be computed by using the preselected positions x_v are subsequently perturbed for generating a random interval having range ε and containing the true distance. The positioning of the true distance in the interval is random because of the random selection of the real value σ in $[0, 1]$. All generated instances contain m distances, each of them represented by a suitable interval. In the vertex ordering that is associated to the instance, every vertex (that is not the first one) has at least one reference vertex.

Table 1 shows the computational experiments with different values for n and for the percentages p_t. The value of ε is set to 0.1 for the generation of all instances. The implementation of the BP1 algorithm allows to choose whether to perform or not the backtracking pruning. For every run, the total number $nsols$ of found solutions and the computational time (in microseconds) are reported. Moreover, in case BP1 performs the backtracking pruning, the number $restarts$ of times the exploration of a branch is restarted is reported as well. Finally, the table contains the values of Δ, computed as

$$\Delta = \frac{range_{\max}^{\text{backp}}}{range_{\max}^{\text{nobackp}}},$$

where $range_{\max}$ is the maximal range of the intervals $[z_v^L, z_v^U]$ composing BP1 solutions; the superscripts "backp" and "nobackp" indicate whether the backtracking pruning was executed or not, respectively, for obtaining the considered solutions. The value of Δ gives the reduction over the largest interval ranges achieved when performing the backtracking pruning.

The experiments took from a few microseconds (10^{-6} s) for the small-sized instances, to some milliseconds (10^{-3} s) for $n \geq 50$, until reaching the threshold of 1 or more seconds for $n \geq 100$. The effect of the backtracking pruning is to increase the total number of solutions, and to provide a final set of mappings where intervals are tighter. In fact, the backtracking pruning ensures that no interval parts can be infeasible, and this leads intervals at lines 4–7 of Algorithm 2 to be split (increased number of solutions) and/or reduced in size (see Δ values) by applying the backtracking pruning. Naturally, the execution of the backtracking pruning, together with the possibility to prune an entire branch and to restart its exploration from a new set of intervals, makes the overall computational cost increase. The percentages p_i, which control the number of reference vertices, strongly influence the total number of found solutions. From all BP1 solutions, it was possible to construct embeddings x by executing Algorithm 3.

Table 1 Experiments on artificially generated instances. The computational time is expressed in microseconds (10^{-6} s)

Instance						BP1 w/out back. pruning		BP1 with back. pruning			
n	m	p_1	p_2	p_3	p_4	nsols	Time	Restarts	nsols	Δ	Time
10	30	0.10	0.10	0.10	0.70	2	23	5	3	0.90	41
10	29	0.10	0.10	0.20	0.60	4	28	238	36	0.83	485
10	28	0.10	0.10	0.30	0.50	2	22	63	6	0.53	100
10	27	0.10	0.20	0.20	0.50	2	27	19	5	0.94	45
10	24	0.10	0.30	0.30	0.30	2	23	10	2	0.83	44
10	23	0.20	0.20	0.30	0.30	2	22	83	4	0.73	161
25	80	0.10	0.10	0.10	0.70	4	63	60	4	0.48	218
25	81	0.10	0.10	0.20	0.60	2	62	74	3	0.86	238
25	75	0.10	0.10	0.30	0.50	2	59	158	3	0.63	431
25	75	0.10	0.20	0.20	0.50	8	102	502	14	0.74	1396
25	65	0.10	0.30	0.30	0.30	2	70	990	35	0.84	2698
50	166	0.10	0.10	0.10	0.70	18	545	19879	378	0.33	91016
50	161	0.10	0.10	0.20	0.60	8	273	1615	23	0.48	7888
50	156	0.10	0.10	0.30	0.50	4	179	2405	152	0.79	12310
50	151	0.10	0.20	0.20	0.50	12	739	10559	156	0.59	43761
100	336	0.10	0.10	0.10	0.70	16	1435	183487	6110	0.67	2888058
100	326	0.10	0.10	0.20	0.60	8	1174	125457	69	0.51	1364581
100	316	0.10	0.10	0.30	0.50	4	1008	116780	64	0.39	1110584
125	420	0.10	0.10	0.10	0.70	24	5306	6020796	1728	0.58	55115773
125	411	0.10	0.10	0.20	0.60	16	2873	1137928	2832	0.38	11807205
150	506	0.10	0.10	0.10	0.70	96	113488	18662764	1920	0.42	273721615

5 Discussion and Conclusion

The DGP in dimension 1 can be solved exactly by the proposed variant of the BP algorithm even when interval distances are given. The BP1 algorithm integrates the so-called "backtracking pruning", which allows to refine all vertex position intervals, so that they only contain feasible positions with respect to the overall set of distances. BP1 can be applied to all instances in dimension 1 for which, in the given vertex order, there exists at least one reference vertex for every vertex of the graph.

It is still an open problem to conceive methods and algorithms for a complete exploration of the solution set of DGPs in dimension $K > 1$. The need to select sample positions from the arcs obtained with the intersections (see Sect. 2) makes the BP algorithm become a heuristic: some of the limitations of this approach were very recently discussed in [10]. When distances are represented by intervals, and $K > 1$, these intersections provide Euclidean objects having dimension 1 [20]. Differently from the intervals obtained during the execution of BP1, however, these Euclidean objects lie in a space that has a higher dimension [12]. This makes the main ideas at the basis of the proposed BP1 not easy to extend to DGPs in any dimensions.

Acknowledgements The author is thankful to the anonymous referees.

References

1. Alipanahi, B., Krislock, N., Ghodsi, A., Wolkowicz, H., Donaldson, L., Li, M.: Determining protein structures from NOESY distance constraints by semidefinite programming. J. Comput. Biol. **20**(4), 296–310 (2013)
2. Alves, R., Cassioli, A., Mucherino, A., Lavor, C., Liberti, L.: The integration of clifford algebra in the i BP algorithm for the DMDGP. In: Proceedings of the International Work-Conference on Bioinformatics and Biomedical Engineering (IWBBIO13), pp. 745–746. Granada, Spain (2013)
3. Alves, R., Cassioli, A., Mucherino, A., Lavor, C., Liberti, L.: Adaptive branching in iBP with clifford algebra. In: Proceedings of Distance Geometry and Applications (DGA13), pp. 65-69. Manaus, Amazonas, Brazil (2013)
4. Billinge, S.J.L., Duxbury, Ph.M., Gonçalves, D.S., Lavor, C., Mucherino, A.: Assigned and unassigned distance geometry: applications to biological molecules and nanostructures. Q. J. Oper. Res. **14**(4), 337–376 (2016)
5. Biswas, P., Lian, T., Wang, T., Ye, Y.: Semidefinite programming based algorithms for sensor network localization. ACM Trans. Sens. Netw. **2**, 188–220 (2006)
6. Crippen, G.M., Havel, T.F.: Distance Geometry and Molecular Conformation. Wiley, New York (1988)
7. Ding, Y., Krislock, N., Qian, J., Wolkowicz, H.: Sensor network localization, Euclidean distance matrix completions, and graph realization. Optim. Eng. **11**(1), 45–66 (2010)
8. Freris, N.M., Graham, S.R., Kumar, P.R.: Fundamental limits on synchronizing clocks over networks. IEEE Trans. Autom. Control **56**(6), 1352–1364 (2010)
9. Gonçalves, D.S., Mucherino, A., Lavor, C.: An adaptive branching scheme for the branch & prune algorithm applied to distance geometry. In: IEEE Conference Proceedings, Federated Conference on Computer Science and Information Systems (FedCSIS14), Workshop on Computational Optimization (WCO14), pp. 463–469. Warsaw, Poland (2014)

10. Gonçalves, D.S., Mucherino, A., Lavor, C., Liberti, L.: Recent advances on the interval distance geometry problem. J. Glob. Optim. (2017). To appear
11. Lavor, C., Liberti, L., Maculan, N., Mucherino, A.: The discretizable molecular distance geometry problem. Comput. Optim. Appl. **52**, 115–146 (2012)
12. Lavor, C., Liberti, L., Mucherino, A.: The interval branch-and-prune algorithm for the discretizable molecular distance geometry problem with inexact distances. J. Glob. Optim. **56**(3), 855–871 (2013)
13. Lavor, C., Alves, R., Figueiredo, W., Petraglia, A., Maculan, N.: Clifford algebra and the discretizable molecular distance geometry problem. Adv. Appl. Clifford Algebr. **25**(4), 925–942 (2015)
14. Liberti, L., Lavor, C., Maculan, N.: A branch-and-prune algorithm for the molecular distance geometry problem. Int. Trans. Oper. Res. **15**, 1–17 (2008)
15. Liberti, L., Lavor, C., Maculan, N., Mucherino, A.: Euclidean distance geometry and applications. SIAM Rev. **56**(1), 3–69 (2014)
16. Mucherino, A., Lavor, C.: The branch and prune algorithm for the molecular distance geometry problem with inexact distances. In: Proceedings of World Academy of Science, Engineering and Technology **58**, International Conference on Bioinformatics and Biomedicine (ICBB09), pp. 349–353. Venice, Italy (2009)
17. Mucherino, A., Lavor, C., Liberti, L.: The discretizable distance geometry problem. Optim. Lett. **6**(8), 1671–1686 (2012)
18. Mucherino, A., Lavor, C., Liberti, L., Maculan, N. (eds.): Distance Geometry: Theory, Methods and Applications, 410 pp. Springer, New York (2013)
19. Mucherino, A., de Freitas, R., Lavor, C.: Distance geometry and applications. Discret. Appl. Math. **197**, 1–144 (2015). Special issue
20. Petitjean, M.: Spheres unions and intersections and some of their applications in molecular modeling. In: [19], pp. 61–83 (2013)
21. Saxe, J.: Embeddability of weighted graphs in k-space is strongly NP-hard. In: Proceedings of 17th Allerton Conference in Communications, Control and Computing, pp. 480–489 (1979)
22. Wang, Z., Zheng, S., Ye, Y., Boyd, S.: Further relaxations of the semidefinite programming approach to sensor network localization. SIAM J. Optim. **19**(2), 655–673 (2008)
23. Wu, Y.-C., Chaudhari, Q., Serpedin, E.: Clock synchronization of wireless sensor networks. IEEE Signal Process. Mag. **28**(1), 124–138 (2011)

Verification of Correctness of Parallel Algorithms in Practice

Jakub Nalepa and Miroslaw Blocho

Abstract Verification of the correctness of parallel algorithms is often omitted in the works from the parallel computation field. In this paper, we discuss in detail how to show that a parallel algorithm is correct. This process involves proving its safety and liveness. In our case study, we prove the correctness of our two parallel algorithms for the NP-hard pickup and delivery problem with time windows. Both algorithms (for minimizing the number of routes and the travel distance) were already shown to be extremely efficient in practice—the implementations were thoroughly examined using the famous Li and Lim's benchmark dataset.

1 Introduction

Parallel algorithms became a standard tool for solving complex optimization problems from many fields, including the computational biology, genomics, text processing, pattern recognition, machine learning, optimization, medical imaging and many others, due to the availability of various parallel architectures. They can be used to quickly traverse the solution space in search of high-quality (and feasible) solutions. Proving the correctness of such parallel approaches is a challenging task (it is much more difficult compared with serial algorithms). However, this issue is very often ignored in works from the parallel computation field.

In this paper, we show how to prove the correctness of a parallel algorithm at hand. Our parallel guided ejection search technique (P–GES) for minimizing the number of trucks in the NP-hard pickup and delivery problem with time windows (PDPTW), along with our parallel memetic algorithm (P–MA, being a hybrid of an evolutionary approach coupled with local-search improvement procedures [17, 34, 42]) for optimizing the travel distance in the PDPTW serve as the case study.

J. Nalepa (✉) · M. Blocho
Institute of Informatics, Silesian University of Technology, Akademicka 16,
44-100 Gliwice, Poland
e-mail: jakub.nalepa@polsl.pl

M. Blocho
e-mail: blochom@gmail.com

© Springer International Publishing AG 2018
S. Fidanova (ed.), *Recent Advances in Computational Optimization*,
Studies in Computational Intelligence 717, DOI 10.1007/978-3-319-59861-1_9

135

We carefully investigate their correctness, and show how to accomplish that in a step-by-step manner that can be easily tailored to other parallel algorithms. In our previous works [9, 27, 32], we experimentally evaluated the Message Passing Interface implementation of P–GES and P–MA. The extensive experimental studies revealed that these algorithms are quite efficient, and they are able to extract very high-quality *feasible* routing schedules (often better than the world's best known solutions at that time) in practice. The analysis of their correctness complements our previous efforts and theoretically proves that P–GES and P–MA are correct parallel algorithms. This work substantially extends our very recent paper on verifying the correctness of such parallel techniques [30].

This paper is organized as follows. Section 2 formulates of the PDPTW. Section 3 reviews the state of the art on solving the PDPTW, and on parallel heuristic and evolutionary algorithms. In Sect. 4, we present the background on verifying the correctness of parallel algorithms. The correctness of our parallel guided search is proven in Sect. 5, whereas the correctness of our parallel memetic algorithm is investigated in Sect. 6. The paper is concluded in Sect. 7, which also presents the directions of our future work.

2   Pickup and Delivery Problem with Time Windows

The PDPTW is an NP-hard discrete optimization problem of serving transportation requests, each being a pair of the pickup and delivery operations. The PDPTW is defined on a directed graph $G = (V, E)$, with a set V of $C + 1$ vertices. The vertices $v_i, i \in \{1, \ldots, C\}$, are the travel points, whereas v_0 denotes the depot (the start and the finish point of each route). A set of edges $E = \{(v_i, v_{i+1}) | v_i, v_{i+1} \in V, v_i \neq v_{i+1}\}$ are the travel connections between each pair of travel points. The travel costs $c_{i,j}$, $i, j \in \{0, 1, \ldots, C\}, i \neq j$, are equal to the distances (they are often given in the Euclidean metric) between the travel points. Each request $h_i, i \in \{0, 1, \ldots, N\}$, where $N = C/2$, is a coupled pair of pickup (P) and delivery (D) customers—these customers are given as p_h and d_h, respectively, where $P \cap D = \emptyset$, and $P \cup D = V \setminus \{v_0\}$. The amount of delivered ($q^d(h_i)$) and picked up ($q^p(h_i)$) goods is defined for each h_i, where $q^d(h_i) = -q^p(h_i)$. Each v_i defines its demand, service time s_i (where $s_0 = 0$), and time window $[e_i, l_i]$ within which the service of this customer should be *started* (it can finish after closing this time window). The fleet (containing K vehicles) is homogenous—the capacity of each truck is equal to Q. Each route r in the solution σ, starts and finishes at the depot, and it is an ordered list of visited travel points.

The PDPTW is a two-objective NP-hard discrete optimization problem—the main objective is to minimize the fleet size K, whereas the secondary one is to optimize the distance $T = \sum_{i=1}^{K} T_i$, where T_i is the distance of the i-th route. Let σ_A and σ_B denote two PDPTW solutions. σ_A is then of a higher quality compared with σ_B (assuming that both solutions are feasible), if $(K(\sigma_A) < K(\sigma_B))$ or $(K(\sigma_A) = K(\sigma_B)$ and $T(\sigma_A) < T(\sigma_B))$.

3 Related Literature

3.1 Solving the Pickup and Delivery with Time Windows

State-of-the-art algorithms for rich routing problems encompass exact and approximate methods [18]. The former algorithms deliver the exact solutions [3, 7, 12, 35], however they are very difficult to apply in practice, because of their unacceptably large execution times (especially in the case of massively large, real-life problem instances). Also, handling the dynamic changes which are very common in many circumstances (e.g., updating the traffic networks to avoid congestion) are not trivial to incorporate in such algorithms [6]. In a majority of exact algorithms, a single objective is considered (e.g., minimizing the travel distance). Exact techniques were discussed in several very interesting and thorough works [4, 13].

Approximation algorithms include construction and improvement heuristics and metaheuristics [1, 2]. The construction (very often referred to as insertion-based) techniques create feasible solutions from scratch by inserting consecutive requests iteratively into the partial solution (not all of the requests are served in a partial solution) [22, 43]. The partial solution encompasses a subset of all transportation requests, therefore is not acceptable and should be expanded to serve all other (currently unserved) transportation requests. On the other hand, improvement heuristics modify an initial solution by applying local search moves—the neighborhood of the initial solution in therefore explored [21, 38]. A number of metaheuristics have been adopted for solving rich vehicle routing problems (VRPs), including various tabu searches [35], variable neighborhood searches [40], greedy randomized adaptive search procedures, population-based [11, 25, 37], and agent-based approaches [19], guided ejection searches [29], simulated annealing [38], and more [24].

In evolutionary algorithms (both sequential and parallel), a population of solutions undergoes the biologically-inspired evolution [39]. The individuals (representing routing schedules) are selected for mating, then they are crossed over and mutated. In memetic algorithms (also known as *hybrid genetic algorithms*), this evolution process is enhanced by the local-search procedures aimed at boosting the quality of already-found solutions (this process, commonly called *education*, often extends the mutation operation). Such evolutionary approaches can be conveniently terminated once the solution of desired quality has been obtained, or the execution time surpassed the imposed time limit. They were shown very efficient and have been applied to solving a plethora of various rich vehicle routing problems [26].

3.2 Parallel Heuristic Algorithms

Parallel heuristic algorithms have been explored for solving a bunch of different optimization problems [15], including various VRPs [19, 26]. Co-operative strategies in such parallel heuristic techniques have been discussed and classified in several taxonomies, with the one presented by Crainic et al. being quite well-established [14]. This taxonomy encompasses three dimensions: the first dimension specifies if the global solving procedure is controlled by a *single process* (1-control—1C) or by a *group of processes* (p-control—pC). These processes may co-operate (in co-operative algorithms[1]) or not (if the processing is in the batch mode). The second dimension reflects the quantity and quality of the information exchanged between the parallel processes, along with the additional knowledge derived from these exchanges (note that the co-operation is useful only if it occurs *on time* and if the data being transferred is *meaningful*). The four classes are defined for this dimension: Rigid (RS), Knowledge Synchronization (KS), Collegial (C) and Knowledge Collegial (KC). The third dimension concerns the diversity of the initial solutions and search strategies: Same Initial Point/Population, Same Search Strategy (SPSS), Same Initial Point/Population, Different Search Strategies (SPDS), Multiple Initial Points/Populations, Same Search Strategies (MPSS), Multiple Initial Points/Populations, Different Search Strategies (MPDS).

As already mentioned, parallel heuristic and metaheuristic algorithms were successfully applied to solve numerous challenging optimization problems from a variety of science and engineering fields, including medical imaging, text categorization, banking, pattern recognition and many more. Such approaches were very intensively explored for solving rich routing problems as well [26, 41], including the PDPTW [9, 27]. The implementations of these algorithms take advantage from the massively-parallel architectures (both with the shared and distributed memory [5], and those equipped with GPU co-processors [20]). These parallel architectures are easily accessible nowadays, and they are relatively easy to exploit efficiently. The parallel techniques are able to deliver extremely high-quality routing schedules in short time (very often the best known solutions of rich vehicles routing problems are retrieved using parallel evolutionary algorithms [26]), even for enormously large problem instances and those with multiple real-life constraints.

4 Verification of the Correctness of Parallel Algorithms

Verification of the correctness of a given sequential algorithm encompasses performing the following steps:

1. Showing that this algorithm will finish (at least one of the termination conditions will be finally met).
2. Showing that this algorithm will give a correct result for any correct set of the input data.

More formally, the correctness may be stated as:

$$\{p\}A\{q\}, \tag{1}$$

---

[1]It is worth mentioning that the co-operation schemes—defining the co-operation topology, frequency and strategies for handling sent and received solutions have tremendous impact on the capabilities and behavior of parallel algorithms, as shown in our previous works: [26, 33].

where A denotes the algorithm (a set of statements executed in the pre-defined order), p is the pre-condition, and q represents the post-condition. The pre-condition specifies which conditions *must* hold for the input data, and the post-condition reflects what *must* be satisfied by the results retrieved using the algorithm A to consider such results feasible. The algorithm is *partially correct* if for any input data satisfying the pre-condition, it gives the correct output data (in accordance with the post-condition) [23] (the input-output relation holds). The algorithm is *totally correct*, if it is partially correct, and—for any input data—it reaches the termination condition and returns the correct output. Therefore, proving the total correctness of a sequential algorithm consists of proving its partial correctness, along with showing that every execution of this algorithm will be finally terminated [8].

In the case of parallel algorithms, proving their correctness includes verifying their *safety* and *liveness*. The algorithm is *safe*, if it never ends up in a forbidden state. To prove the safety of a parallel algorithm, we need to show that:

1. The parallel algorithm is *partially correct*.
2. There are *no deadlocks* (i.e., the processes do not wait for the infinite amount of time for each other to continue the execution).
3. The processes can safely access the shared resources (*mutual exclusion*).

The liveness property of a parallel algorithm is satisfied, if it can be proven that a certain desired condition will eventually happen during the execution [16, 36]. In the case of message-passing techniques—as shown in [8]—it is important to show that the messages are properly sent and received (no matter if the communication is synchronous or asynchronous).

If all of the above-mentioned properties of a parallel algorithm are proven, then this algorithm is *correct*.

5   Correctness of the Parallel Guided Ejection Search for the PDPTW

The baseline (sequential) version of the GES was proposed in [24], and later enhanced and parallelized in our very recent works [9, 27, 29]. According to the taxonomy presented in Sect. 3.2, P–GES is of the pC/C/MPSS type (p-Control, Collegial, Multiple Initial Points, Same Search Strategies). In Sect. 5.1, we give the overview of this parallel GES, whereas its correctness is verified in Sect. 5.2.

5.1   Algorithm Outline

In P–GES, which is an improvement parallel heuristic technique, p processes are executed in parallel (Algorithm 1, line 1). The initial feasible solution σ contains the number of routes which is equal to the number of transportation requests—each

---

Algorithm 1 A parallel algorithm to minimize K (P–GES).

1: **for** $P_i \leftarrow P_1$ **to** P_p **do in parallel**
2:     $\tau_{last} \leftarrow \tau_{curr}$;
3:     Create an initial solution σ;
4:     *finished* $\leftarrow$ **false**;
5:     **while not** *finished* **do**
6:         Save the current feasible solution;
7:         Put requests from a random route r into EP;
8:         Set penalty counters $p[i] \leftarrow 1(i = 1, 2, \ldots, N)$;
9:         **while** (EP $\neq \emptyset$) **and** (**not** *finished*) **do**
10:             Select and remove request h_{in} from EP;
11:             **if** $S_{in}^{fe}(h_{in}, \sigma) \neq \emptyset$ **then**
12:                 $\sigma \leftarrow$ random $\sigma' \in S_{in}^{fe}(h_{in}, \sigma)$;
13:             **else**
14:                 $\sigma \leftarrow$ **Squeeze**(h_{in}, σ);
15:             **end if**
16:             **if** h_{in} is not inserted into σ **then**
17:                 $p[h_{in}] \leftarrow p[h_{in}] + 1$;
18:                 **for** $k \leftarrow 1$ **to** k_m **do**
19:                     Get $S_{ej}^{fe}(h_{in}, \sigma)$ with min. $\mathscr{P}_{sum}$;
20:                     **if** $S_{ej}^{fe}(h_{in}, \sigma) \neq \emptyset$ **then**
21:                         $\sigma \leftarrow$ random $\sigma' \in S_{ej}^{fe}(h_{in}, \sigma)$;
22:                         Add $(h_{out}^{(1)}, h_{out}^{(2)}, \ldots, h_{out}^{(k)})$ to EP;
23:                         **break**;
24:                     **end if**
25:                 **end for**
26:             **end if**
27:             $\sigma \leftarrow$ **Perturb**(σ);
28:             **if** $\tau_{curr} \geq \tau_{last} + \tau_{coop}$ **then**
29:                 *finished* $\leftarrow$ **Cooperate**(σ);
30:                 $\tau_{last} \leftarrow \tau_{curr}$;
31:             **end if**
32:         **end while**
33:         **if** EP $\neq \emptyset$ **then**
34:             Backtrack to previous feasible solution;
35:         **end if**
36:     **end while**
37:     Get the best solution σ_{best};
38: **end for**

---

request is served by a separate truck (line 3). Then, the number of vehicles in is consecutively decreased until the total computation time exceeds the imposed time limit τ_M (lines 5–36), or the desired number of routes has been obtained (it is possible to determine the minimal number of trucks of a given capacity that are necessary to serve all transportation requests feasibly [26]).

A random route r is removed from the solution σ, and the excluded requests are pushed into the *ejection pool* (EP), which stores those transportation requests that have been removed from the schedule, and which remain currently unserved (the solution becomes the partial solution; line 7). The penalty counters (denoted as p's) are reset (line 8). These counters reflect the difficulty of re-inserting a given request back into the partial solution (the higher counter value, the more difficult is to re-insert the corresponding request into σ).

If the EP contains unserved transportation requests (lines 9–32), then a single request h_{in} is popped from the EP at the time (line 10), and the attempts to re-insert it into the partial solution are undertaken. If there exist any feasible insertion positions (i.e., those which do not violate the constraints) for this request (the set of such positions $S_{in}^{fe}(h_{in}, \sigma)$ is not empty), then a random position is drawn (line 12). Otherwise, the request is inserted into σ infeasibly (it violates the constraints), and the feasibility of the solution is being restored in the *squeezing* procedure (line 14). The solution penalty is quantified using the penalty function given as:

$$\mathcal{F}_p(\sigma) = \mathcal{F}_c(\sigma) + \mathcal{F}_{tw}(\sigma), \tag{2}$$

where $\mathcal{F}_c(\sigma)$ and $\mathcal{F}_{tw}(\sigma)$ are the sum of capacity exceeds in σ, and the sum of the time windows violations. The squeeze function (presented in Algorithm 2) is aimed at decreasing the value of the penalty function until it reaches zero (thus, the solution becomes feasible). This is a steepest-descent local search procedure, in which the set $S^{inf}(h_{in}, r, \sigma_t)$ of infeasible solutions is created (considering the insertion of the transportation request at hand), and the solution with the minimal value of the penalty function is selected. This process continues until the feasibility is restored, or it is impossible to retrieve a feasible solution (in this case, the solution is backtracked to the initial state).

Algorithm 2 Squeezing an infeasible (possibly partial) solution σ.

1: **function** SQUEEZE(h_{in}, σ)
2:    $\sigma_t \leftarrow \sigma' \in S^{inf}(h_{in}, \sigma)$ such that $\mathcal{F}_p(\sigma')$ is minimum;
3:    **while** ($\mathcal{F}_p(\sigma_t) \neq 0$) **do**
4:        Randomly choose an infeasible route r in σ_t;
5:        Find $\sigma'' \in S^{inf}(h_{in}, r, \sigma_t)$ with min. $\mathcal{F}_p(\sigma'')$;
6:        **if** $\mathcal{F}_p(\sigma'') < \mathcal{F}_p(\sigma_t)$ **then**
7:            $\sigma_t \leftarrow \sigma''$;
8:        **else**
9:            **break**;
10:       **end if**
11:   **end while**
12:   **if** $\mathcal{F}_p(\sigma_t) = 0$ **then**
13:       **return** σ_t;
14:   **else**
15:       **return** σ;
16:   **end if**
17: **end function**

If the squeeze procedure fails, the penalty counter of the appropriate request $(p[h_{in}])$ is updated (Algorithm 1, line 17), and other requests are ejected from σ (up to k_m requests are ejected; lines 18–25) to insert h_{in}. The set $S_{ej}^{fe}(h_{in}, \sigma)$ is formed, and it encompasses the solutions with various combinations of ejected requests (the h_{in} request is inserted to this solution on various positions). Finally, the solution σ'— with the minimal sum of the penalty counters is selected from $S_{ej}^{fe}(h_{in}, \sigma)$ (line 21). The ejected requests are added to the EP (line 22). The solution σ is finally perturbed by the local search procedure, in which I feasible local moves (out-relocate and out-exchange) are executed (line 27). This process is visualized in Algorithm 3.

Algorithm 3 Perturbing a feasible (possibly partial) solution σ.

1: **function** PERTURB(σ)
2:     $\sigma_t \leftarrow \sigma$;
3:     **for** $i \leftarrow 1$ **do** I
4:         Find σ' through local search moves on σ_t;
5:         **if** σ' is feasible **then**
6:             $\sigma_t \leftarrow \sigma'$;
7:         **end if**
8:     **end for**
9:     **return** σ_t;
10: **end function**

The parallel processes in P–GES co-operate periodically every τ_{coop} seconds (Algorithm 1, line 29) using the asynchronous co-operation scheme. In our previous works [26, 27], we investigated a number of co-operation schemes (as already mentioned, they define the co-operation topology, frequency, and the strategies for handling emigrants/immigrants—the solutions being sent/received in the co-operation phase) and showed, that a proper selection of such scheme has a crucial influence on the algorithm capabilities and behavior.

In P–GES, the master process (P_1) controls the execution time of the parallel algorithm—the signals from P_1 to either continue or terminate the execution are transferred in each co-operation phase. Eventually, all solutions from all processes are gathered in P_1, and the best solution σ_{best} is retrieved—this is the final solution delivered by P–GES (line 37).

More details on P–GES can be found in our previous works [9, 27]. These papers include the in-depth analysis of the Message Passing Interface (MPI) implementation of the algorithm, and discuss the experimental results retrieved for famous Li and Lim's benchmark sets (these tests reflect various real-life scheduling scenarios, e.g., different positions of the travel points, tightness of time windows, and different capacities of the available trucks).

5.2 Proving the Correctness of P–GES

The input data passed to P–GES include:

- p ($p \geq 1$)—the number of parallel processes. If $p = 1$, then P–GES becomes a serial algorithm, and its certain components are disabled (e.g., the co-operation between the processes).
- $K_d \geq 0$—the desired number of trucks serving the requests. If $K_d = 0$, then the best feasible solution found using P–GES is returned (i.e., there is no "desired" number of routes, however K should be as minimum as possible).
- τ_{MAX}—the maximum execution time (in seconds) of P–GES.
- τ_{coop}—the co-operation frequency (in seconds).
- k_m ($k_m \geq 1$)—the maximum number of requests that can be ejected from a (possibly partial) solution while inserting a request popped from the EP.
- I ($I \geq 0$)—the number of local search moves applied to perturb a solution.
- Test instance—the definition of the test instance. It specifies the number of transportation requests, the positions of the travel points, their time windows, service times, and demands (either pickup or delivery), and the maximum capacity of trucks. Real-life problems may encompass travel points which are clustered, randomly scattered around the map, or combine both (i.e., there are some customer clusters, but lots of them are random). The problem instances belonging to the Li and Lim's benchmark set perfectly mimic these scenarios.[2]

The solution retrieved using P–GES *must* satisfy all the constraints discussed in Sect. 2. Therefore, this solution must be *feasible* (otherwise, the routing schedule is incorrect and cannot be accepted).

As mentioned in Sect. 5.1, P–GES starts with an initial feasible solution (therefore, the constraints are not violated at the beginning of the algorithm execution), in which every transportation request is served in a separate route. Then, the attempts to reduce the fleet size are performed, until the execution reaches the termination condition (Algorithm 1, line 5)—one random route is analyzed at any time. The current (best) solution is stored (line 6). If removing the selected random route fails, then the partial solution is backtracked to this state (line 34), hence the routing schedule remains feasible.

Once the ejected customers are pushed into the EP, the solution becomes a partial feasible schedule (no constraints are violated). Then, the EP transportation requests are put back into this partial solution—first, using the feasible insertion positions (if any). In this case, the feasibility is not violated, and the next request from the EP is popped for insertion. However, if there are no feasible insertion positions available, then the infeasible solution is created (in which the request has been re-inserted back infeasibly), and it is further processed with the squeeze procedure. This squeezing retrieves either the feasible solution (if it is possible to restore the correctness of σ), or

---

[2]For the instance definitions, and the world's best solutions see: https://www.sintef.no/projectweb/top/pdptw/li-lim-benchmark/. Reference date: January 30, 2017.

backtracks it to the state before this squeezing has been called. In the latter case, other transportation requests are ejected to restore the feasibility of the partial solution—it finally becomes feasible. Perturbing a feasible (potentially partial) schedule can deteriorate its quality, however it cannot cause violating the constraints—after calling this procedure, the solution remains feasible. If the EP is empty, then the feasible solution—with the decreased fleet size—becomes the next solution, which is to be processed in the next algorithm iteration. Therefore, the PDPTW solution obtained using P–GES is eventually *always feasible*.

The co-operation of parallel processes cannot affect the feasibility of the solutions. Depending on the co-operation scheme, the receiving process may e.g., replace its own solution with the immigrant (if the immigrant is of a higher quality). This analysis shows that P–GES is *partially correct*—assuming that the input data are correct, it always retrieves a feasible PDPTW solution.

P–GES may be terminated if (i) a solution of a desired quality (i.e., with the desired number of routes, K_d) is found, or (ii) if the maximum execution time has been exceeded. In the former case, the final solution may be retrieved either by the master process (which also controls the execution of other processes, and may send the termination request), or by any other process (not a master process). If the master retrieves this solution, then it sends the termination requests to other processes (thus, one co-operation phase is enough to stop the execution). However, if another process ends up with the desired solution, then two co-operation phases are necessary—first, this process sends its best solution to the master, and then the master sends the termination request to other processes. In either case, P–GES *finally reaches its stopping condition*.

P–GES is a distributed algorithm (there are no shared resources). The co-operation is asynchronous, and the execution (i.e., optimization of the solution run by a given process) interleaves with the send/receive operations. The order of send/receive operations matters—they are executed in an appropriate order depending on the process type (either the master on non-master). Additionally, receiving data is acknowledged by the receiving process during the co-operation (the status of this acknowledgement is periodically checked by the sending process). Since there are no deadlocks and shared resources in P–GES, its *safety is proven*. The same reasoning may be used to prove the liveness of the algorithm. Since only the master process can force other processes to stop, the situation in which a given process sends to or waits for a message from the process that has already been terminated is not possible. This shows the liveness property of P–GES.

The above-presented investigation revealed that all of the conditions imposed on the parallel algorithms to be correct are fulfilled by P–GES—for the correct input data (e.g., assuming that the test instance is *solvable*). Therefore, **P–GES is a correct parallel algorithm**. □

6 Correctness of the Parallel Memetic Algorithm for the PDPTW

In the parallel memetic algorithm (abbreviated as P–MA) for minimizing the travel distance in the PDPTW, a population of feasible solutions undergoes the memetic optimization. This is the island-model parallel evolutionary algorithm, in which every process (also referred to as the *island*) evolves its own population, and the islands communicate periodically to exchange the best solutions found so far, and to guide the search effectively (P–MA is therefore of the pC/C/MPSS type according to the taxonomy presented in Sect. 3.2). The initial population can be generated using any route minimization algorithm. In our work, we exploit the parallel GES discussed at length in the previous section (however, it can be very easily replaced with any other technique, without influencing the parallel memetic approach). P–MA is discussed in Sect. 6.1, and its correctness is proven in Sect. 6.2.

6.1 Algorithm Outline

In P–MA (Algorithm 4), the number of trucks is minimized first (line 1), and then the initial population of size N_{pop} containing only feasible solutions is generated (lines 3–5). Here, each individual (being a PDPTW solution) has the same number of routes m. Afterwards, each (out of p) island performs the memetic optimization of its population (lines 6–26).

During the evolution, the pairs of parents are selected for crossover (line 9)—here, we utilize the AB-selection scheme, in which every individual from the population serves exactly once as the first parent, and once as the second one. If the parents are of the same structure (note that the solution of the PDPTW can be represented as a directed graph), then one individual is perturbed (line 12) (we exploit the same perturb method as in P–GES). This parent structure is modified because crossing over the individuals representing exactly the same solution would most likely not result in retrieving higher-quality children. The parents are then crossed over (line 15) using the improved selective route exchange crossover (SREX), proposed in [24], and very recently substantially improved in our work [10] (applying this operator cannot lead to obtaining not feasible routing schedules). After generating N_{ch} children (in order to fully exploit each pair of mating parents), these offspring solutions are educated. The education process is visualized in Algorithm 5—it resembles the perturb operation (a number of local-search edge-exchange and relocation moves are performed), however it *cannot* lead to lower-quality neighboring solutions (only the solutions with lower T values are accepted in the course of this procedure). Once the child individuals are educated, the best one replaces the first parent and survives to the next generation (Algorithm 4, line 22). It is worth noting that the elitism is implicitly applied here—the best solution always survives.

The parallel islands periodically co-operate (see Sect. 5.1 for details—we use the same co-operation strategy). Finally, the termination conditions are verified (line 25).

Algorithm 4 Parallel memetic algorithm for the PDPTW (P–MA).

1: $\sigma_1 \leftarrow$ **MinimizeRoutes**();   ▷ P–GES
2: $m \leftarrow$ routes count of σ_1;
3: **for** $P_i \leftarrow P_1$ **to** P_p **do in parallel**
4:     Generate the population of N_{pop} feasible solutions;
5: **end for**
6: **for** $P_i \leftarrow P_1$ **to** P_p **do in parallel**
7:     $done \leftarrow$ **false**;
8:     **while not** $done$ **do**
9:         Determine N_{pop} random pairs (σ_A^p, σ_B^p);
10:         **for** all pairs (σ_A^p, σ_B^p) **do**
11:             **if** $\sigma_A^p = \sigma_B^p$ **then**
12:                 $\sigma_A^p \leftarrow$ **Perturb**(σ_A^p);
13:             **end if**
14:             $\sigma_{best}^c \leftarrow \sigma_A^p$;
15:             $\{\sigma_1^c, \sigma_2^c, \ldots, \sigma_{N_{ch}}^c\} \leftarrow$ execute N_{ch} times **Crossover**(σ_A^p, σ_B^p);
16:             **for** $i \leftarrow 1$ **to** N_{ch} **do**
17:                 $\sigma_i^c \leftarrow$ **Educate**(σ_i^c);
18:                 **if** $T(\sigma_i^c) < T(\sigma_{best}^c)$ **then**
19:                     $\sigma_{best}^c \leftarrow \sigma_i^c$;
20:                 **end if**
21:             **end for**
22:             $\sigma_A^p \leftarrow \sigma_{best}^c$;
23:         **end for**
24:         **Co-operate**();
25:         $done \leftarrow$ VerifyStoppingCondition();
26:     **end while**
27: **end for**
28: **return** best solution σ_{best} across all populations;

P–MA may be stopped if (i) the solution of desired quality has been already retrieved, (ii) the maximum execution time has been surpassed, (iii) the maximum number of generations have been processed, or (iv) there is no significant improvement in the solutions quality between several consecutive generations (it indicates that the memetic algorithm converged and the best individual will likely not be improved any further). Finally, the best solution (across all the islands) is returned (line 28). More details on the parallel memetic algorithm, along with the implementation details (P–MA was implemented as a distributed algorithm using MPI) and the experimental results can be found in [10, 26, 28, 32].

6.2 Proving the Correctness of P–MA

The input data passed to P–GES include:

- p ($p \geq 1$)—the number of parallel processes. If $p = 1$, then P–MA becomes a serial algorithm, and its certain components are disabled (e.g., the co-operation between the processes).

Algorithm 5 Educating a feasible solution σ.

1: **function** EDUCATE(σ)
2:     $\sigma_t \leftarrow \sigma$;
3:     **for** $i \leftarrow 1$ **do** I
4:         Find σ' through local search moves on σ_t;
5:         **if** σ' is feasible **and** $T(\sigma')<T(\sigma)$ **then**
6:             $\sigma_t \leftarrow \sigma'$;
7:         **end if**
8:     **end for**
9:     **return** σ_t;
10: **end function**

- N_{pop} ($N_{pop} \geq 2$)—the size of the population evolved in each island.
- N_{ch} ($N_{ch} \geq 1$)—the number of children generated for each pair of mating parents in the crossover process.
- τ_{MAX}^{K}—the maximum execution time of P–GES (during the route minimization phase).
- τ_{MAX}^{Pop}—the maximum execution time of P–GES (during the population generation phase).
- τ_{MAX}—the maximum execution time (in seconds) of P–MA.
- τ_{coop}—the co-operation frequency (in seconds).
- Test instance—the definition of the test instance.

As already mentioned, the initial steps of P–MA involves executing P–GES to minimize the number of trucks, and to generate the initial populations for each island. The correctness of our parallel GES has been proven in Sect. 5.2. Similarly, the routing schedules retrieved using P–MA *must* satisfy all the constraints imposed on the PDPTW solutions presented in Sect. 3.

The initial populations encompass only *feasible* PDPTW solutions (no constraints are violated). Then, these populations undergo the evolution to optimize the travel distance. The perturb method (executed for some parental solution in Algorithm 4, line 12), cannot violate the constraints, therefore the solution remains feasible. Then, N_{ch} are created in the recombination process. As presented in [10], the improved SREX retrieves only feasible offspring solutions (if obtaining such schedules is possible). Afterwards, those feasible children are educated (line 17; similarly, this procedure cannot cause violating the constraints). The best child solution survives, and replaces one parent in the post-selection process (note that if there are no children of higher quality than the parent, this parent is not removed from the population). Therefore, the T optimization process will always lead to obtaining feasible PDPTW solutions. As presented in the previous section, the co-operation between processes does not affect the feasibility of the transferred individuals. Hence, P–MA is *partially correct* (for any correct input data, it *always retrieves a feasible PDPTW solution*).

There are the time limits imposed on the route minimization phase (τ_{MAX}^{K}), and the population generation phase (τ_{MAX}^{Pop}). If the execution time surpasses these limits, the parallel GES is terminated and the current solutions (with the current number of trucks) are considered final. If this happens for the population generation phase, then the already-retrieved routing schedules are copied and perturbed to meet the desired population size (N_{pop}). P–MA (i.e., its T optimization phase) may be terminated if (i) the solution of desired quality has been already retrieved, (ii) the maximum execution time has been surpassed, (iii) the maximum number of generations have been processed, or (iv) there is no significant improvement in the solutions quality between several consecutive generations. In either case, the master process sends the termination request to other processes during the next co-operation phase (according to the selected co-operation scheme). Then, those processes terminate and the final solution is retrieved. Hence, *P–MA will finally reach its termination condition.*

The proofs for the safety and liveness of P–MA are analogous to those presented for P–GES. Therefore, **P–MA is a correct parallel algorithm**, and the entire framework, consisting of the parallel guided ejection search for minimizing the number of trucks in the PDPTW, and the parallel memetic algorithm for optimizing the travel distance is **fully correct**. □

7 Conclusions and Outlook

In this paper, we analyzed the correctness of our (i) parallel guided ejection search algorithm for minimizing the number of trucks in the PDPTW, and our (ii) parallel memetic algorithm to optimize the travel distance in the PDPTW (those approaches are combined into a single parallel framework for solving this NP-hard discrete optimization problem). We proved that both algorithms are correct—this involved showing their liveness and safety. The investigation served as an extensive case study for showing how to prove the correctness of parallel algorithms. Our approach may be easily tailored for proving the correctness of other parallel algorithms, especially those tackling complex discrete optimization problems.

Our current research is focused on designing a generic optimization framework which will allow for solving other complex transportation problems. This will be possible by implementing a mechanism for adding the appropriate constraints. Also, we work on machine learning based methods which are aimed at retrieving the best possible algorithm variant (e.g., with the desired parameter values) for a given problem instance (in our previous works we showed that improperly tuned parameters can easily jeopardize the search capabilities of metaheuristic methods for solving rich routing problems [28, 31]).

Acknowledgements This research was supported by the National Science Centre under research Grant No. DEC-2013/09/N/ST6/03461, and performed using the infrastructure supported by the POIG.02.03.01-24-099/13 grant: "GeCONiI—Upper Silesian Center for Computational Science and Engineering", and the Intel CPU and Xeon Phi platforms provided by the MICLAB project No. POIG.02.03.00.24-093/13.

References

1. Akeb, H., Bouchakhchoukha, A., Hifi, M.: A beam search based algorithm for the capacitated vehicle routing problem with time windows. In: Proceedings of the 2013 Federated Conference on Computer Science and Information Systems, Kraków, Poland, September 8–11, 2013, pp. 329–336 (2013). http://ieeexplore.ieee.org/xpl/articleDetails.jsp?arnumber=6644021
2. Akeb, H., Bouchakhchoukha, A., Hifi, M.: A three-stage heuristic for the capacitated vehicle routing problem with time windows. Recent Advances in Computational Optimization: Results of the Workshop on Computational Optimization WCO 2013, FedCSIS 2013, pp. 1–19. Springer International Publishing, Cham (2015). doi:10.1007/978-3-319-12631-9_1
3. Baldacci, R., Bartolini, E., Mingozzi, A.: An exact algorithm for the pickup and delivery problem with time windows. Oper. Res. **59**(2), 414–426 (2011). doi:10.1287/opre.1100.0881
4. Baldacci, R., Mingozzi, A., Roberti, R.: Recent exact algorithms for solving the vehicle routing problem under capacity and time window constraints. Eur. J. Oper. Res. **218**(1), 1–6 (2012)
5. Banos, R., Ortega, J., Gil, C., de Toro, F., Montoya, M.G.: Analysis of OpenMP and MPI implementations of meta-heuristics for vehicle routing problems. Appl. Soft Comput. **43**, 262–275 (2016). doi:10.1016/j.asoc.2016.02.035, http://www.sciencedirect.com/science/article/pii/S1568494616300862
6. Bernay, B., Deleplanque, S., Quilliot, A.: Routing on dynamic networks: GRASP versus genetic. In: Proceedings of the 2014 Federated Conference on Computer Science and Information Systems, Warsaw, Poland, September 7–10, 2014, pp. 487–492 (2014). doi:10.15439/2014F52
7. Bettinelli, A., Ceselli, A., Righini, G.: A branch-and-price algorithm for the multi-depot heterogeneous-fleet pickup and delivery problem with soft time windows. Math. Program. Comput. **6**(2), 171–197 (2014). doi:10.1007/s12532-014-0064-0
8. Blocho, M.: A parallel memetic algorithm for the vehicle routing problem with time windows. Ph.D. thesis, Silesian University of Technology (2013). (in Polish)
9. Blocho, M., Nalepa, J.: A parallel algorithm for minimizing the fleet size in the pickup and delivery problem with time windows. In: Proceedings of 22nd European MPI Users' Group Meeting, EuroMPI '15, pp. 15:1–15:2. ACM, New York, USA (2015). doi:10.1145/2802658.2802673
10. Blocho, M., Nalepa, J.: LCS-Based selective route exchange crossover for the pickup and delivery problem with time windows. In: Hu, B., López-Ibáñez, M. (eds.) Evolutionary Computation in Combinatorial Optimization: 17th European Conference, EvoCOP 2017, Amsterdam, The Netherlands, April 19–21, 2017, Proceedings, pp. 124–140. Springer International Publishing, Cham (2017). doi:10.1007/978-3-319-55453-2_9
11. Cherkesly, M., Desaulniers, G., Laporte, G.: A population-based metaheuristic for the pickup and delivery problem with time windows and LIFO loading. Comput. Oper. Res. **62**, 23–35 (2015). doi:10.1016/j.cor.2015.04.002, http://www.sciencedirect.com/science/article/pii/S0305054815000829
12. Cordeau, J.F.: A branch-and-cut algorithm for the dial-a-ride problem. Oper. Res. **54**(3), 573–586 (2006). doi:10.1287/opre.1060.0283
13. Cordeau, J.F., Laporte, G., Ropke, S.: Recent models and algorithms for one-to-one pickup and delivery problems. In: The Vehicle Routing Problem: Latest Advances and New Challenges, pp. 327–357. Springer, Boston, MA (2008). doi:10.1007/978-0-387-77778-8_15
14. Crainic, T.G., Nourredine, H.: Parallel meta-heuristics applications. In: Gendreau, M., Potvin, J.Y. (eds.) Parallel Metaheuristics: A New Class of Algorithms, pp. 447–494. Wiley (2005). doi:10.1007/978-1-4419-1665-5_17
15. Crainic, T.G., Toulouse, M.: Parallel meta-heuristics. In: Gendreau, M., Potvin, J.Y. (eds.) Handbook of Metaheuristics, International Series in Operations Research & Management Science, vol. 146, pp. 497–541. Springer, US (2010). doi:10.1007/978-1-4419-1665-5_17
16. Czech, Z.: Introduction to Parallel Computing. PWN (2013)

17. Fidanova, S., Alba, E., Molina, G.: Memetic simulated annealing for the GPS surveying problem. In: 4th International Conference on Numerical Analysis and Its Applications, NAA 2008, Lozenetz, Bulgaria, June 16–20, 2008. Revised Selected Papers, pp. 281–288 (2008). doi:10.1007/978-3-642-00464-3_30

18. Grandinetti, L., Guerriero, F., Pezzella, F., Pisacane, O.: The multi-objective multi-vehicle pickup and delivery problem with time windows. Soc. Beh. Sc. **111**, 203–212 (2014). doi:10.1016/j.sbspro.2014.01.053, http://www.sciencedirect.com/science/article/pii/S1877042814000548

19. Kalina, P., Vokrinek, J.: Parallel solver for vehicle routing and pickup and delivery problems with time windows based on agent negotiation. In: IEEE SMC, pp. 1558–1563 (2012)

20. Karpinski, M., Pacut, M.: Multi-GPU parallel memetic algorithm for capacitated vehicle routing problem. CoRR **abs/1401.5216** (2014). arXiv:1401.5216

21. Li, H., Lim, A.: A metaheuristic for the pickup and delivery problem with time windows. In: Proc. IEEE ICTAI, pp. 160–167 (2001). doi:10.1109/ICTAI.2001.974461

22. Lu, Q., Dessouky, M.M.: A new insertion-based construction heuristic for solving the pickup and delivery problem with time windows. Eur. J. Oper. Res. **175**(2), 672–687 (2006). doi:10.1016/j.ejor.2005.05.012, http://www.sciencedirect.com/science/article/pii/S0377221705004698

23. Manna, Z.: Mathematical theory of partial correctness. J. Comput. Syst. Sci. **5**(3), 239–253 (1971). doi:10.1016/S0022-0000(71)80035-1

24. Nagata, Y., Kobayashi, S.: Guided ejection search for the pickup and delivery problem with time windows. In: Proceedings of EvoCOP, *LNCS*, vol. 6022, pp. 202–213. Springer (2010). doi:10.1007/978-3-642-12139-5_18

25. Nagata, Y., Kobayashi, S.: A memetic algorithm for the pickup and delivery problem with time windows using selective route exchange crossover. In: Proceedings of PPSN XI pp. 536–545. Springer, Heidelberg (2010). doi:10.1007/978-3-642-15844-5_54

26. Nalepa, J., Blocho, M.: Co-operation in the parallel memetic algorithm. Int. J. Parallel Program. **43**(5), 812–839 (2015). doi:10.1007/s10766-014-0343-4

27. Nalepa, J., Blocho, M.: A parallel algorithm with the search space partition for the pickup and delivery with time windows. In: Proceedings of IEEE 3PGCIC, pp. 92–99 (2015). doi:10.1109/3PGCIC.2015.12

28. Nalepa, J., Blocho, M.: Adaptive memetic algorithm for minimizing distance in the vehicle routing problem with time windows. Soft Comput. **20**(6), 2309–2327 (2016). doi:10.1007/s00500-015-1642-4

29. Nalepa, J., Blocho, M.: Enhanced guided ejection search for the pickup and delivery problem with time windows. In: Intelligent Information and Database Systems: Proceedings of 8th Asian Conference, ACIIDS 2016, pp. 388–398. Springer, Heidelberg (2016). doi:10.1007/978-3-662-49381-6_37

30. Nalepa, J., Blocho, M.: Is your parallel algorithm correct? In: Position Papers of the 2016 Federated Conference on Computer Science and Information Systems, FedCSIS 2016, Gdańsk, Poland, September 11–14, 2016, pp. 87–93 (2016). doi:10.15439/2016F554

31. Nalepa, J., Blocho, M.: Adaptive guided ejection search for pickup and delivery with time windows. J. Intell. Fuzzy Syst. **32**(2), 1547–1559 (2017). doi:10.3233/JIFS-169149

32. Nalepa, J., Blocho, M.: 2017 25th Euromicro International Conference on Parallel, Distributed and Network-based Processing (PDP). A Parallel Memetic Algorithm for the Pickup and Delivery Problem with Time Windows, pp. 1–8. March (2017). doi:10.1109/PDP.2017.75

33. Nalepa, J., Blocho, M.: Temporally adaptive co-operation schemes. In: Xhafa, F., Barolli, L., Amato, F. (eds.) Advances on P2P, Parallel, Grid, Cloud and Internet Computing: Proceedings of the 11th International Conference on P2P, Parallel, Grid, Cloud and Internet Computing (3PGCIC–2016) November 5–7, 2016, Asan, Korea, pp. 145–156. Springer International Publishing, Cham (2017). doi:10.1007/978-3-319-49109-7_14

34. Nalepa, J., Kawulok, M.: Adaptive memetic algorithm enhanced with data geometry analysis to select training data for SVMs. Neurocomputing **185**, 113–132 (2016). doi:10.1016/j.neucom.2015.12.046, http://www.sciencedirect.com/science/article/pii/S0925231215019839

35. Nanry, W.P., Barnes, J.W.: Solving the pickup and delivery problem with time windows using reactive tabu search. Transp. Res. **34**(2), 107–121 (2000). doi:10.1016/S0191-2615(99)00016-8

36. Owicki, S., Lamport, L.: Proving liveness properties of concurrent programs. ACM Trans. Program. Lang. Syst. **4**(3), 455–495 (1982)

37. Pankratz, G.: A grouping genetic algorithm for the pickup and delivery problem with time windows. OR Spectr. **27**(1), 21–41 (2005)

38. Parragh, S.N., Doerner, K.F., Hartl, R.F.: A survey on pickup and delivery problems. Journal fur Betriebswirtschaft **58**(1), 21–51 (2008)

39. Roeva, O., Fidanova, S., Paprzycki, M.: Population size influence on the genetic and ant algorithms performance in case of cultivation process modeling. In: Recent Advances in Computational Optimization - Results of the Workshop on Computational Optimization WCO 2013 [FedCSIS 2013, Kraków, Poland], pp. 107–120 (2013). doi:10.1007/978-3-319-12631-9_7

40. Ropke, S., Pisinger, D.: An adaptive large neighborhood search heuristic for the pickup and delivery problem with time windows. Transp. Sci. **40**(4), 455–472 (2006). doi:10.1287/trsc.1050.0135

41. Senarclens de Grancy, G., Reimann, M.: Evaluating two new heuristics for constructing customer clusters in a vrptw with multiple service workers. Cent. Eur. J. Oper. Res. **23**(2), 479–500 (2015). doi:10.1007/s10100-014-0373-4

42. Siminski, K.: Memetic neuro-fuzzy system with differential optimisation. In: Kozielski, S., Mrozek, D., Kasprowski, P., Małysiak-Mrozek, B., Kostrzewa, D. (eds.) Beyond Databases, Architectures and Structures. Advanced Technologies for Data Mining and Knowledge Discovery: 12th International Conference, BDAS 2016, Ustroń, Poland, May 31 - June 3, 2016, Proceedings, pp. 135–145. Springer International Publishing, Cham (2016)

43. Zhou, C., Tan, Y., Liao, L., Liu, Y.: Solving the multi-vehicle pick-up and delivery problem with time widows by new construction heuristic. Proc. IEEE CISDA **2**, 1035–1042 (2006). doi:10.1109/ISDA.2006.253754

Application of Fuzzy Cognitive Maps with Evolutionary Learning Algorithm to Model Decision Support Systems Based on Real-Life and Historical Data

Katarzyna Poczeta, Łukasz Kubuś, Alexander Yastrebov
and Elpiniki I. Papageorgiou

Abstract Fuzzy cognitive map (FCM) is a universal tool for modeling dynamic decision support systems. It can be constructed by the experts or learned based on historical data. FCM models learned from data are denser than those created by humans. We developed an evolutionary learning approach for fuzzy cognitive maps based on density and system performance indicators. It allows to select only the most significant connections between concepts and receive the structure more similar to the FCMs initialized by experts. This paper is devoted to the application of the developed approach to model decision support systems with the use of real-life and historical data.

1 Introduction

Fuzzy cognitive map (FCM) is a directed graph for representing causal reasoning [9]. Nodes are variable concepts important for the analyzed problem and links are causal connections. FCM can be used for modeling decision support systems [12], e.g. for political decision making [3], artificial emotions forecasting [18] or prediction of work of complex systems [20].

K. Poczeta (✉) · Ł. Kubuś · A. Yastrebov
Kielce University of Technology, al. Tysiąclecia Państwa Polskiego 7, 25-314
Kielce, Poland
e-mail: k.piotrowska@tu.kielce.pl

Ł. Kubuś
e-mail: l.kubus@tu.kielce.pl

A. Yastrebov
e-mail: a.jastriebow@tu.kielce.pl

E.I. Papageorgiou
Technological Educational Institute (T.E.I.) of Central Greece, 3rd Km Old National
Road Lamia-Athens, 35100 Lamia, Greece
e-mail: epapageorgiou@teiste.gr; elpiniki.papageorgiou@uhasselt.be

E.I. Papageorgiou
Faculty of Business Economics, Hasselt University, Campus Diepenbeek Agoralaan
Gebouw D, 3590 Diepenbeek, Belgium

© Springer International Publishing AG 2018
S. Fidanova (ed.), *Recent Advances in Computational Optimization*,
Studies in Computational Intelligence 717, DOI 10.1007/978-3-319-59861-1_10

Fuzzy cognitive map can be initialized based on expert knowledge. Experts select the concepts of the map and determine the weights of the connections between them (connection matrix). The second way to build the FCM model are supervised [8] and evolutionary learning algorithms [1, 5, 12, 21, 23] that allow to determine the connection matrix based on available data.

The resulting matrices are much denser than the models created by the humans. The density of the FCMs developed by experts is usually in the range of 30–40% [22]. They choose only the most significant connections between concepts. FCMs with the smaller density are more readable for humans. Developing the learning algorithms that allow to build models in a manner more similar to human reasoning is an important part of research related to the fuzzy cognitive maps. A sparse real-coded genetic algorithm was proposed to utilize the density of the FCM model [22]. In [15], a structure optimization genetic algorithm was introduced. A multi-objective evolutionary algorithm for learning maps with varying densities was analyzed in [6].

In [11], we proposed a new evolutionary learning approach for fuzzy cognitive maps learning based on density and system performance indicators (SPI) analysis. System performance indicators introduced by Borisov [4] and Silov [19] allow to analyze reliability of the FCM model and determine the total (direct and indirect) influence between concepts. In the proposed approach, the evaluation of the candidate FCMs is based on data error, connection matrix density and the total influence between concepts. The obtained results based on the synthetic and real-life data generated from the reference matrices provided by experts proved that the developed approach allows to receive structure of the FCM model more similar to the reference object keeping the similar level of data error. In [16] the proposed evolutionary learning algorithm has been effectively used to construct the economic decision support system using historical data.

This paper is devoted to further analysis of the application of fuzzy cognitive maps with the developed evolutionary algorithm based on SPI in modeling decision support systems. The aim of the analysis is to approximate the real-life and historical data. The comparison of the developed approach with the standard one based on data error and the approach based on density was done. The learning process was accomplished with the use of Elite Genetic Algorithm (EGA) and Individually Directional Evolutionary Algorithm (IDEA) [10].

Section 2 presents fuzzy cognitive maps. Section 3 describes system performance indicators. The developed evolutionary algorithm for fuzzy cognitive maps learning is described in Sect. 4. In Sect. 5, the results of experiments based on real-life [17] and historical [7] data are presented. Section 6 contains the conclusion of the paper.

2   Fuzzy Cognitive Maps

Fuzzy cognitive map is a directed graph in the form [9]:

$$< X, W > \tag{1}$$

where $X = [X_1, \ldots, X_n]^T$ is the set of the concepts, W is the connection matrix describing weights of the connections, $w_{j,i}$ is the weight of the direct influence between the j-th concept and the i-th concept, taking on the values from the range $[-1, 1]$. A positive weight of the connection $w_{j,i}$ means that X_j causally increases X_i. A negative weight of the connection $w_{j,i}$ means that X_j causally decreases X_i.

Fuzzy cognitive map can be used for modeling behavior of dynamic systems. The state of the FCM model is determined by the values of the concepts at the t-th iteration. The simulation of the FCM behavior requires an initial state vector. Next, the values of the concepts can be calculated according to the selected dynamic model. Simulations show the effect of the changes in the states of the map and can be used in a what-if analysis [2]. In the paper a popular dynamic model was used [21]:

$$X_i(t+1) = F\left(\sum_{j=1, j\neq i}^{n} w_{j,i} \cdot X_j(t)\right) \tag{2}$$

where $X_i(t)$ is the value of the i-th concept at the t-th iteration, $i = 1, 2, \ldots, n$, n is the number of concepts, t is discreet time, $t = 0, 1, 2, \ldots, T$. Transformation function $F(x)$ normalizes values of the concepts to a proper range. A logistic function is most often used [21, 22]:

$$F(x) = \frac{1}{1 + e^{-cx}} \tag{3}$$

where c is a parameter determined experimentally, $c > 0$.

3 System Performance Indicators

System performance indicators allow to evaluate the structure of the FCM model e.g. by analysis the influence between concepts. Connection matrix W describes the direct influence between concepts. The total influence $p_{j,i}$ between concepts means the maximum direct or indirect influence between concepts. To determine this system performance indicator the total causal effect path between concepts can be calculated [4, 19].

Algorithm for determining SPI contains the following steps [4, 19]:

1. First, connection matrix W with positive and negative direct relationships between concepts passes to matrix R size $2n \times 2n$ with positive relationships as follows:

$$\begin{aligned} &\text{if } w_{j,i} > 0 \\ &\text{then } r_{2j-1,2i-1} = w_{j,i}, \, r_{2j,2i} = w_{j,i} \\ &\text{if } w_{j,i} < 0 \\ &\text{then } r_{2j-1,2i} = -w_{j,i}, \, r_{2j,2i-1} = -w_{j,i} \end{aligned} \tag{4}$$

2. Next, operation of transitive closure of the matrix R is used:

$$R^* = R \vee R^2 \vee R^3 \vee \ldots \tag{5}$$

where $\vee$ means maximum operation, R^k is calculated in accordance with the max-product composition:

$$R^k = R^{k-1} \circ R \tag{6}$$

3. Elements of the matrix R^* are transformed into matrix V as follows:

$$\begin{aligned}
v_{j,i} &= max(r_{2j-1,2i-1}, r_{2j,2i}) \\
v'_{j,i} &= -max(r_{2j-1,2i}, r_{2j-1,2i})
\end{aligned} \tag{7}$$

4. On the basis of the matrix V the total (direct and indirect) influence between the j-th concept and the i-th concept is calculated:

$$\begin{aligned}
&\text{for } v_{j,i} \neq v'_{j,i} \\
p_{j,i} &= sign(v_{j,i} + v'_{j,i})max(|v_{j,i}|, |v'_{j,i}|)
\end{aligned} \tag{8}$$

where $p_{j,i}$ takes value in the range of $[-1, 1]$.

5. The total influence between concepts can be used to determine the impact of the j-th concept on the system (9) and the impact of the system on the i-th concept (10):

$$\overrightarrow{P_j} = \frac{1}{n} \sum_{i=1}^{n} p_{j,i} \tag{9}$$

$$\overleftarrow{P_i} = \frac{1}{n} \sum_{j=1}^{n} p_{j,i}. \tag{10}$$

4 Evolutionary Learning Algorithm Based on SPI

Evolutionary algorithms (like RCGA or EGA) can be applied to learn the FCM model (determine the weights of the connections between concepts) based on the available historical data. Each individual in the population is represented by a floating-point vector [21]:

$$W' = [w_{1,2}, \ldots, w_{1,n}, w_{2,1}, w_{2,3}, \ldots, w_{2,n}, \ldots, w_{n,n-1}]^T \tag{11}$$

where $w_{j,i}$ is the weight of the connection between the j-th and the i-th concept.

The individuals are decoded into the candidate FCMs and the response of every model are calculated based on the learning initial state vectors. The aim of the standard evolutionary algorithms for fuzzy cognitive maps learning is to minimize a total difference between the normalized historical data and the model response (data error), described as follows [21]:

$$TE = \sum_{p=1}^{P} \sum_{t=1}^{T} \sum_{i=1}^{n} |Z_i^p(t) - X_i^p(t)| \tag{12}$$

where $t = 0, 1, 2, \ldots, T$, T is the learning record length, $Z_i^p(t)$ is the reference value of the i-th concept at iteration t for the p-th learning record, $X_i^p(t)$ is the value of the i-th concept at iteration t of the candidate FCM started from the p-th initial state vector, $p = 1, 2, \ldots, P$, P is the number of the learning records.

Fuzzy cognitive maps learned with the use of methods based on data error properly perform the task of the input data approximation. However, the resulting connection matrices are denser than those initialized by experts [22]. Density of the FCM model can be expressed as a ratio of the number of non-zero weights and number of all possible non-zero weights according to the formula:

$$density = \frac{w_{non-zero}}{n^2 - n} \tag{13}$$

where $w_{non-zero}$ is the number of non-zero weights $w_{j,i}$, n is the number of the concepts.

To solve the problem of density the extensions of the standard evolutionary algorithms based on density analysis were proposed:

- sparse real-coded genetic algorithm [22],
- structure optimization genetic algorithm [15],
- multi-objective evolutionary algorithm for learning FCM models with varying densities [6].

In [11], we introduced a new evolutionary approach for fuzzy cognitive maps learning that allow to determine the weights of the connections in the way similar to human reasoning. It is based on the analysis of data error, density and total influence between concepts in order to select only the most significant connections. Experiments performed with the use of synthetic and real-life data (generated from the reference FCMs) confirm that the resulting models are more similar to the reference systems.

The developed algorithm has the following objectives [11]:

- to minimize the data error (12),
- to minimize the FCM model density (13),
- to maximize the ratio of the number of significant total influences between concepts to the number of all possible influences described as follows:

$$PChRR = \frac{p_{relevant}}{n^2} \tag{14}$$

where $p_{relevant}$ is the number of total influences with the absolute value greater than 0.5 ($|p_{j,i}| > 0.5$).

To obtain some compromise between data error, density and significance of the influences between concepts, weighting method for determine the objective function was used [14]:

$$Error = a_1 \cdot TE + a_2 \cdot density \cdot TE +$$
$$+a_3 \cdot (1 - PChRR) \cdot TE \tag{15}$$

where a_1, a_2, a_3 are parameters that meet the following condition:

$$\sum_{i=1}^{3} a_i = 1 \tag{16}$$

Density and total influence objectives are multiplied by the total error in order to lie in the same range.

Each candidate FCM is evaluated with the use of the following fitness function:

$$fitness(Error) = -Error \tag{17}$$

The developed approach consists of the following steps [11]:

STEP 1. Initialize random population.

Random initial population is generated and evaluated with the use of the fitness function. Each generated individual has density greater or equal to 20% and lower or equal to 50%.

STEP 2. Check stop condition.

If the number of iterations is greater than $iteration_{max}$ then stop the learning process.

STEP 3. Use evolutionary algorithm to generate new population.

In the simulation analysis of the proposed approach Elite Genetic Algorithm and Individually Directed Evolutionary Algorithm were used [10].

STEP 4. Analyze population.

The values from $[-0.05, 0.05]$ are rounded down to 0 as suggested in [21]. The total influence between concepts $p_{j,i}$ is calculated. The weight value $w_{j,i}$ is rounded down to 0 if the value of $p_{j,i}$ is in the interval $[-b, b]$, where b is a parameter selected experimentally. Additionally, density is checked. Go to **STEP 2**.

STEP 5. Choose the best individual and test it.

4.1 Elite Genetic Algorithm

Elite Genetic Algorithm (EGA) [13] uses floating-point encoding as Real-Coded Genetic Algorithm (RCGA) and elite strategy. In the first step, random initial population is generated and evaluated. Next, temporary population T^t is created from current base population P^t by proportionate selection (roulette-wheel selection) with dynamic linear scaling of fitness function. Individuals of temporary population T^t are modified by Uniform Crossover operator and Non-Uniform Mutation operator [13]. The Uniform Crossover uses a fixed mixing ratio between two parents called *exchange probability* and usually it is equal to 0.5. Non-Uniform Mutation (NUM) operator keeps the population from stagnating in the early stages and decreases the range of mutation in later stages of evolution. In the last step, the temporary population T^t becomes new base population P^{t+1} after evaluation of the temporary population individuals.

4.2 Individually Directed Evolutionary Algorithm

Individually Directional Evolutionary Algorithm (IDEA) [10] uses floating-point encoding as EGA expanded by the additional mutation direction vector (DV). The DV is used by the mutation operation and correction of mutation direction process in post-selection stage. Random initial population is generated and evaluated as for EGA. Next, the roulette-wheel selection with dynamic linear scaling of fitness function is used to create temporary population T^t. Next, Directional Non-Uniform Mutation operator (DNUM) is used to create the second population T'^t based on the temporary population T^t [10]. Individuals of temporary population T'^t are evaluated and the next base population is created with the use of post-selection. The individual of temporary population T^t is compared to the corresponding individual from temporary population T'^t. Better individual is selected for the next base population. If the mutated individual is worse than the corresponding individual, the corresponding element of directional vector of the primary individual is corrected (correction of mutation direction).

5   Experiments

The aim of the experiments is to build the decision support systems that allows to approximate the input data and analyze the influence between concepts. The comparative analysis of the developed approach with the standard one based on data error and the approach based on density was done.

5.1 Evaluation Criteria

To evaluate the resulting FCM models, the following criteria were calculated:

1. Similarity between the input learning data and the data generated by the FCM candidate [21]:

$$initial_{error} = \frac{1}{P \cdot T \cdot n} \sum_{t=1}^{T} \sum_{p=1}^{P} \sum_{i=1}^{n} |Z_i^p(t) - X_i^p(t)| \qquad (18)$$

where $t = 0, 1, 2, \ldots, T$, T is the learning record length, $Z_i^p(t)$ is the reference value of the i-th concept at iteration t for the p-th learning record, $X_i^p(t)$ is the value of the i-th concept at iteration t of the candidate FCM started from the p-th initial state vector, $p = 1, 2, \ldots, P$, P is the number of the learning records.

2. Generalization capabilities of the candidate FCM (similarity between the input testing data and the data generated by the FCM candidate) [21]:

$$behavior_{error} = \frac{1}{P \cdot T \cdot n} \sum_{p=1}^{P} \sum_{t=1}^{T} \sum_{i=1}^{n} |Z_i^p(t) - X_i^p(t)| \qquad (19)$$

where $t = 0, 1, 2, \ldots, T$, T is the testing record length, $Z_i^p(t)$ is the reference value of the i-th concept at iteration t for the p-th testing record, $X_i^p(t)$ is the value of the i-th concept at iteration t of the candidate FCM started from the p-th initial state vector, $p = 1, 2, \ldots, P$, P is the number of the testing records.

3. Similarity between the candidate FCM structure and the real-life model [22]:

$$weights_{error} = \frac{1}{n^2} \sum_{i=1}^{n} \sum_{j=1}^{n} |w'_{j,i} - w_{j,i}| \qquad (20)$$

where $w_{j,i}$ is the weight of the connection between the j-th and the i-th concept in the FCM candidate and $w'_{j,i}$ is the weight of the connection between the j-th and the i-th concept in the input model.

5.2 Datasets

To analyze the performance of the developed evolutionary algorithm for fuzzy cognitive maps learning real-life [17] and historical [7] data were used.

Real-Life Data

Real-life data were obtained based on the fuzzy cognitive map for a mobile payment system project with the following concepts [17]:

- X_1 – your phone is always with you,
- X_2 – remote control of everyday things,
- X_3 – independence of time and place,
- X_4 – flexibility,
- X_5 – multiple mobile cash accounts,
- X_6 – convenience,

- X_7 – interface easy to use,
- X_8 – efficiency,
- X_9 – direct debiting from account,
- X_{10} – comfort,
- X_{11} – bank commission from network operator for each transaction,
- X_{12} – economy.

Figure 1 shows the analyzed decision support system based on fuzzy cognitive map. Table 1 presents its connection matrix.

The input data for the learning and testing process were generated starting from the random initial vectors. The FCM models were learned using 10 learning state vectors ($P = 10$). The resulting models were tested on the basis of the 10 following testing state vectors ($P = 10$) and evaluated with the use of criteria (18)–(20).

Historical Data

Historical data were obtained on the basis of the dataset that contains the hourly averaged responses from an array of 5 metal oxide chemical sensors embedded in an Air Quality Chemical Multisensor Device deployed on the field in an Italian city [7]. The following concepts were determined based on data attributes:

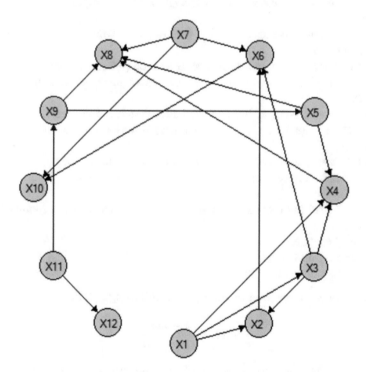

Fig. 1 Fuzzy cognitive map for the mobile payment system project

Table 1 Connection matrix for the mobile payment system project

$w_{j,i}$	X_1	X_2	X_3	X_4	X_5	X_6	X_7	X_8	X_9	X_{10}	X_{11}	X_{12}
X_1	0	0.71	0.83	0.78	0	0	0	0	0	0	0	0
X_2	0	0	0	0	0	0.76	0	0	0	0	0	0
X_3	0	0.74	0	0.87	0	0.79	0	0	0	0	0	0
X_4	0	0	0	0	0	0	0	0.88	0	0	0	0
X_5	0	0	0·	0.65	0	0	0	0.61	0	0	0	0
X_6	0	0	0	0	0	0	0	0	0	0.79	0	0
X_7	0	0	0	0	0	0.76	0	0.89	0	0.85	0	0
X_8	0	0	0	0	0	0	0	0	0	0	0	0
X_9	0	0	0	0	0.62	0	0	0.74	0	0	0	0
X_{10}	0	0	0	0	0	0	0	0	0	0	0	0
X_{11}	0	0	0	0	0	0	0	0	0.7	0	0	−0.56
X_{12}	0	0	0	0	0	0	0	0	0	0	0	0

- X_1 – true hourly averaged concentration CO (reference analyzer),
- X_2 – PT08.S1 (tin oxide) hourly averaged sensor response (nominally CO targeted),
- X_3 – true hourly averaged overall Non Metanic HydroCarbons concentration (reference analyzer),
- X_4 – true hourly averaged Benzene concentration (reference analyzer),
- X_5 – PT08.S2 (titania) hourly averaged sensor response (nominally NMHC targeted),
- X_6 – true hourly averaged NOx concentration (reference analyzer),
- X_7 – PT08.S3 (tungsten oxide) hourly averaged sensor response (nominally NOx targeted),
- X_8 – true hourly averaged NO2 concentration (reference analyzer),
- X_9 – PT08.S4 (tungsten oxide) hourly averaged sensor response (nominally NO2 targeted),
- X_{10} – PT08.S5 (indium oxide) hourly averaged sensor response (nominally O3 targeted),
- X_{11} – temperature,
- X_{12} – relative humidity,
- X_{13} – AH absolute humidity.

The data were normalized according to the following equation:

$$f(x) = \frac{x - min}{max - min} \,, \tag{21}$$

where x is an input numeric value, *min* is the minimum of the dataset, *max* is the maximum of the dataset.

The FCM models were learned using historical normalized data from 5 consecutive days ($P = 5$). The resulting models were tested on the basis of the data from 5 following days ($P = 5$) and evaluated with the use of criteria (18) and (19).

5.3   Learning Parameters

Learning parameters were selected by trial and error. In this paper selected results of the analysis are presented. The following parameters were used for the EGA algorithm:

- selection method: roulette wheel selection with linear scaling,
- recombination method: uniform crossover,
- crossover probability: 0.75,
- mutation method: non-uniform mutation,
- mutation probability: 0.02,
- population size: 10,
- number of elite individuals: 2,
- maximum number of iterations: 500,
- parameter b: 0.1.

The following parameters were used for the IDEA algorithm:

- selection method: roulette wheel selection with linear scaling,
- mutation method: directed non-uniform mutation,
- mutation probability: $\frac{1}{n^2-n}$,
- population size: 10,
- maximum number of iterations: 500,
- parameter b: 0.1.

5.4   Results with Real-Life Data

Table 2 summarizes the results of the experiments with real-life data obtained for the standard approach (STD), the approach based on density (DEN) and the proposed approach (SPI). 10 experiments were performed for every set of the learning parameters and the average values (Avg) and standard deviations (Std) were calculated.

The obtained results show that the developed approach allows to approximate the real-life data with satisfactory accuracy similar to the standard approach and the approach based on density and receive the structure of the FCM model more similar to the real-life system. The lowest average values of the weights error were obtained for the developed approach based on SPI ($weights_{error} = 0.238 \pm 0.349$ for the IDEA learning algorithm and $weights_{error} = 0.224 \pm 0.350$ for the EGA learning algorithm). Moreover, appropriate selection of the parameters a_1, a_2, a_3 allows

Table 2 Experimental results with real-life data

Approach	Method	Parameters	$initial_{error}$ Avg ± Std	$behavior_{error}$ Avg ± Std	$weights_{error}$ Avg ± Std
STD	IDEA	$a_1 = 1\ a_2 = a_3 = 0$	0.017 ± 0.002	0.017 ± 0.001	0.498 ± 0.352
STD	EGA	$a_1 = 1\ a_2 = a_3 = 0$	0.016 ± 0.001	0.018 ± 0.001	0.500 ± 0.363
DEN	IDEA	$a_1 = 0.9\ a_2 = 0.1\ a_3 = 0$	0.011 ± 0.001	0.012 ± 0.002	0.287 ± 0.351
		$a_1 = 0.8\ a_2 = 0.2\ a_3 = 0$	0.025 ± 0.047	0.026 ± 0.046	**0.263** ± 0.353
		$a_1 = 0.7\ a_2 = 0.3\ a_3 = 0$	0.010 ± 0.001	0.011 ± 0.001	0.266 ± 0.346
		$a_1 = 0.6\ a_2 = 0.4\ a_3 = 0$	0.010 ± 0.001	0.011 ± 0.002	0.271 ± 0.358
		$a_1 = 0.5\ a_2 = 0.5\ a_3 = 0$	0.011 ± 0.002	0.012 ± 0.002	0.294 ± 0.372
DEN	EGA	$a_1 = 0.9\ a_2 = 0.1\ a_3 = 0$	0.018 ± 0.003	0.020 ± 0.003	0.331 ± 0.383
		$a_1 = 0.8\ a_2 = 0.2\ a_3 = 0$	0.018 ± 0.004	0.020 ± 0.003	0.330 ± 0.387
		$a_1 = 0.7\ a_2 = 0.3\ a_3 = 0$	0.019 ± 0.002	0.019 ± 0.002	0.331 ± 0.382
		$a_1 = 0.6\ a_2 = 0.4\ a_3 = 0$	0.018 ± 0.001	0.019 ± 0.002	**0.318** ± 0.375
		$a_1 = 0.5\ a_2 = 0.5\ a_3 = 0$	0.019 ± 0.003	0.021 ± 0.004	0.333 ± 0.388
SPI	IDEA	$a_1 = 0.8\ a_2 = 0.1\ a_3 = 0.1$	0.015 ± 0.009	0.016 ± 0.009	0.270 ± 0.350
		$a_1 = 0.1\ a_2 = 0.8\ a_3 = 0.1$	0.019 ± 0.016	0.021 ± 0.016	0.247 ± 0.347
		$a_1 = 0.1\ a_2 = 0.1\ a_3 = 0.8$	0.014 ± 0.003	0.015 ± 0.002	0.348 ± 0.398
		$a_1 = 0.7\ a_2 = 0.0\ a_3 = 0.3$	0.026 ± 0.038	0.027 ± 0.038	**0.238** ± 0.349
		$a_1 = 0.4\ a_2 = 0.3\ a_3 = 0.3$	0.014 ± 0.007	0.015 ± 0.007	0.277 ± 0.360
		$a_1 = 0.3\ a_2 = 0.4\ a_3 = 0.3$	0.024 ± 0.021	0.025 ± 0.021	0.280 ± 0.368
		$a_1 = 0.3\ a_2 = 0.3\ a_3 = 0.4$	0.013 ± 0.006	0.014 ± 0.005	0.290 ± 0.373
		$a_1 = 0.6\ a_2 = 0.0\ a_3 = 0.4$	0.013 ± 0.007	0.015 ± 0.008	0.301 ± 0.369
SPI	EGA	$a_1 = 0.8\ a_2 = 0.1\ a_3 = 0.1$	0.044 ± 0.022	0.045 ± 0.022	0.275 ± 0.369
		$a_1 = 0.1\ a_2 = 0.8\ a_3 = 0.1$	0.061 ± 0.027	0.062 ± 0.027	**0.224** ± 0.350
		$a_1 = 0.1\ a_2 = 0.1\ a_3 = 0.8$	0.030 ± 0.011	0.032 ± 0.012	0.356 ± 0.408
		$a_1 = 0.7\ a_2 = 0.0\ a_3 = 0.3$	0.038 ± 0.024	0.039 ± 0.023	0.313 ± 0.382
		$a_1 = 0.4\ a_2 = 0.3\ a_3 = 0.3$	0.041 ± 0.022	0.042 ± 0.022	0.320 ± 0.396
		$a_1 = 0.3\ a_2 = 0.4\ a_3 = 0.3$	0.032 ± 0.013	0.033 ± 0.012	0.323 ± 0.392
		$a_1 = 0.3\ a_2 = 0.3\ a_3 = 0.4$	0.024 ± 0.010	0.025 ± 0.010	0.323 ± 0.388
		$a_1 = 0.6\ a_2 = 0.0\ a_3 = 0.4$	0.034 ± 0.013	0.035 ± 0.013	0.322 ± 0.392

to obtain some compromise between data errors ($initial_{error}$ and $behavior_{error}$) and similarity between the candidate FCM structure and the real-life model ($weights_{error}$).

To illustrate the proposed approach the best solutions for the selected simulations are presented below. Table 3 presents the resulting connection matrix obtained for the standard approach ($weights_{error} = 0.489$). Connection matrix for the approach based on density ($weights_{error} = 0.285$) is shown in Table 4. Table 5 presents the connection matrix obtained for the proposed approach ($weights_{error} = 0.218$). Table 6 contains impact of the j-th concept on the system (9) and impact of the system on the i-th concept (10) for the real-life model and the obtained FCM models.

The FCM models obtained for the developed approach and method based on density are more readable and easier to interpret than the map obtained with the

Table 3 Connection matrix for the FCM learned with the use of the standard approach

$w_{j,i}$	X_1	X_2	X_3	X_4	X_5	X_6	X_7	X_8	X_9	X_{10}	X_{11}	X_{12}
X_1	0	-0.23	-0.5	0.28	-0.88	0.67	0.81	0.23	0.51	0.16	-0.08	-0.66
X_2	-0.67	0	-0.8	0.36	0.38	0.87	-0.87	0.77	0.13	-0.9	-0.88	-1
X_3	-0.03	0.09	0	-0.2	0.12	0.1	0.84	0.53	-0.8	0.81	0.38	0.45
X_4	0.71	0.17	0.52	0	-0.91	-0.74	-0.14	-0.58	0.09	-0.23	0.37	0.34
X_5	0.36	-0.22	-0.64	-0.17	0	-0.25	-0.16	0.98	0.58	-0.14	-0.05	0.01
X_6	-0.5	-0.84	-0.37	-0.62	0.78	0	0.11	-0.19	0.64	-0.17	0.61	0.77
X_7	1	0.91	0.02	0.98	0.38	-0.96	0	0.25	-0.76	0.87	-0.92	0.72
X_8	-0.35	0.56	0.21	-0.13	-0.7	0.85	0.4	0	-0.89	0.85	0.46	0.04
X_9	-0.66	0.59	0.5	0.59	0.61	0.63	0.73	0.81	0	-0.21	0.35	-0.2
X_{10}	-0.39	-0.29	0.74	0.93	0.71	-0.23	-0.69	0.06	0.41	0	-0.36	-0.75
X_{11}	0.72	0.18	0.67	-0.5	-0.23	0.92	-0.46	0.79	0.9	1	0	-0.26
X_{12}	0.38	0.85	0.42	1	0.41	0.63	-0.58	-0.26	-0.2	-0.13	-0.26	0

Table 4 Connection matrix for the FCM learned with the use of the approach based on density

$w_{j,i}$	X_1	X_2	X_3	X_4	X_5	X_6	X_7	X_8	X_9	X_{10}	X_{11}	X_{12}
X_1	0	0.58	0.7	0.08	0.8	−0.06	0.6	0.92	0	0.91	0	0
X_2	0	0	0	0	−0.45	0	−0.18	0.96	0	0	0	0
X_3	0	0	0	0.27	0	0.67	0	0	0	0	0	0.09
X_4	0	−0.06	0.15	0	0.21	0.38	0	−0.19	0	0	−0.48	−0.11
X_5	0.96	0.28	−0.75	0	0	0.31	−0.63	0	−0.18	0	0	0
X_6	−0.56	0	0.18	0	−0.42	0	0	−0.28	0.08	0.73	−0.07	−0.37
X_7	0	0.98	0	0	−0.31	0	0	0.92	0.56	0	−0.1	0
X_8	0	−0.48	0	0	0.77	0	0	0	0	0	0.51	0
X_9	0	0.4	0	0	0	0	0	0	0	0	0	0
X_{10}	0	0	0	0.87	0	0.4	0.29	0.36	0	0	0.1	0
X_{11}	−0.31	0	0	0.93	0	0	0	0.86	0	0	0	0
X_{12}	0	0	0.63	0	0	0.67	0	0	0.24	0	0	0

Table 5 Connection matrix for the FCM learned with the use of the proposed approach

$w_{j,i}$	X_1	X_2	X_3	X_4	X_5	X_6	X_7	X_8	X_9	X_{10}	X_{11}	X_{12}
X_1	0	0	0.4	0.89	0	0	0	0	−0.28	0.92	0	0
X_2	0	0	0	0	0	0	0	0	0	0	0	0
X_3	0	0	0	0	0	0	0	0	0	0	0	0
X_4	0	0	0	0	0	0	0	1	0	0	0	0
X_5	0	1	0	−0.27	0	0.79	0	0	0.17	0.36	0	−0.67
X_6	0	0.65	0.89	0	−0.2	0	0	1	0.7	−0.48	0	0.07
X_7	0	0	0	0	0	0	0	0	0	0	0	0
X_8	0	0	0.33	0	0	0	0	0	0	0	0	0
X_9	0	0	−1	0	0	0	0	0	0	0	0	0
X_{10}	0	0	−0.12	0.26	0.75	0.97	0	0	0	0	0	0
X_{11}	0.78	−0.18	0	0.89	0	0.67	0	0	−0.18	0.48	0	0.12
X_{12}	−0.93	−0.44	0	0.84	0	−0.23	0	0.98	−0.07	1	0	0

Table 6 System performance indicators obtained for the analyzed approaches

	X_1	X_2	X_3	X_4	X_5	X_6	X_7	X_8	X_9	X_{10}	X_{11}	X_{12}
Real-life $\overrightarrow{P_j}$	0.348	0.113	0.316	0.073	0.105	0.066	0.208	0.0	0.147	0.0	0.115	0.0
Real-life $\overleftarrow{P_i}$	0.0	0.121	0.069	0.249	0.088	0.247	0.0	0.424	0.058	0.282	0.0	−0.047
STD $\overrightarrow{P_j}$	−0.12	−0.44	−0.1	−0.16	0.05	0.38	−0.28	0.01	0.03	−0.23	0.29	−0.37
STD $\overleftarrow{P_i}$	0.16	0.0	−0.1	0.57	−0.53	−0.23	0.02	−0.27	−0.13	0.01	−0.17	−0.27
DEN $\overrightarrow{P_j}$	0.6	0.18	−0.06	−0.1	0.17	−0.15	0.36	0.16	0.12	0.2	0.16	0.04
DEN $\overleftarrow{P_i}$	0.29	−0.02	−0.2	0.53	0.31	0.12	−0.19	0.36	−0.01	0.52	0.04	−0.07
SPI $\overrightarrow{P_j}$	0.53	0.0	0.0	0.11	0.17	0.18	0.0	0.03	−0.08	0.34	0.48	0.45
SPI $\overleftarrow{P_i}$	0.1	0.36	0.37	0.15	0.16	0.32	0.0	0.54	0.31	0.08	0.0	−0.19

standard learning algorithm based only on data error. The developed approach allows to keep only the most significant connections between concepts and receive the structure most similar to the real-life model.

5.5 Results with Historical Data

Table 7 summarizes the results of the experiments with historical data obtained for the standard approach (STD), the approach based on density (DEN) and the proposed approach (SPI).

Table 7 Experimental results with historical data

Approach	Method	Parameters	$initial_{error}$ Avg ± Std	$behavior_{error}$ Avg ± Std
STD	IDEA	$a_1 = 1\ a_2 = a_3 = 0$	0.090 ± 0.001	0.099 ± 0.001
STD	EGA	$a_1 = 1\ a_2 = a_3 = 0$	0.090 ± 0.001	0.099 ± 0.001
DEN	IDEA	$a_1 = 0.9\ a_2 = 0.1\ a_3 = 0$	0.090 ± 0.001	0.099 ± 0.001
		$a_1 = 0.8\ a_2 = 0.2\ a_3 = 0$	0.090 ± 0.001	0.099 ± 0.001
		$a_1 = 0.7\ a_2 = 0.3\ a_3 = 0$	0.090 ± 0.001	0.099 ± 0.001
		$a_1 = 0.6\ a_2 = 0.4\ a_3 = 0$	0.090 ± 0.001	0.099 ± 0.001
		$a_1 = 0.5\ a_2 = 0.5\ a_3 = 0$	0.090 ± 0.001	0.099 ± 0.001
DEN	EGA	$a_1 = 0.9\ a_2 = 0.1\ a_3 = 0$	0.095 ± 0.003	0.102 ± 0.003
		$a_1 = 0.8\ a_2 = 0.2\ a_3 = 0$	0.093 ± 0.002	0.102 ± 0.002
		$a_1 = 0.7\ a_2 = 0.3\ a_3 = 0$	0.094 ± 0.002	0.103 ± 0.003
		$a_1 = 0.6\ a_2 = 0.4\ a_3 = 0$	0.095 ± 0.004	0.103 ± 0.003
		$a_1 = 0.5\ a_2 = 0.5\ a_3 = 0$	0.094 ± 0.003	0.101 ± 0.003
SPI	IDEA	$a_1 = 0.8\ a_2 = 0.1\ a_3 = 0.1$	0.094 ± 0.008	0.103 ± 0.008
		$a_1 = 0.1\ a_2 = 0.8\ a_3 = 0.1$	0.108 ± 0.019	0.113 ± 0.019
		$a_1 = 0.1\ a_2 = 0.1\ a_3 = 0.8$	0.091 ± 0.001	0.099 ± 0.001
		$a_1 = 0.7\ a_2 = 0.0\ a_3 = 0.3$	0.090 ± 0.001	0.099 ± 0.002
		$a_1 = 0.4\ a_2 = 0.3\ a_3 = 0.3$	0.094 ± 0.011	0.102 ± 0.011
		$a_1 = 0.3\ a_2 = 0.4\ a_3 = 0.3$	0.094 ± 0.009	0.102 ± 0.008
		$a_1 = 0.3\ a_2 = 0.3\ a_3 = 0.4$	0.090 ± 0.001	0.099 ± 0.001
		$a_1 = 0.6\ a_2 = 0.0\ a_3 = 0.4$	0.091 ± 0.002	0.099 ± 0.001
SPI	EGA	$a_1 = 0.8\ a_2 = 0.1\ a_3 = 0.1$	0.109 ± 0.018	0.114 ± 0.016
		$a_1 = 0.1\ a_2 = 0.8\ a_3 = 0.1$	0.175 ± 0.050	0.174 ± 0.047
		$a_1 = 0.1\ a_2 = 0.1\ a_3 = 0.8$	0.098 ± 0.008	0.105 ± 0.005
		$a_1 = 0.7\ a_2 = 0.0\ a_3 = 0.3$	0.107 ± 0.021	0.114 ± 0.021
		$a_1 = 0.4\ a_2 = 0.3\ a_3 = 0.3$	0.100 ± 0.011	0.106 ± 0.010
		$a_1 = 0.3\ a_2 = 0.4\ a_3 = 0.3$	0.100 ± 0.014	0.107 ± 0.012
		$a_1 = 0.3\ a_2 = 0.3\ a_3 = 0.4$	0.095 ± 0.004	0.101 ± 0.007
		$a_1 = 0.6\ a_2 = 0.0\ a_3 = 0.4$	0.098 ± 0.004	0.105 ± 0.003

Table 8 Connection matrix for the FCM learned with the use of the standard approach

$w_{j,i}$	X_1	X_2	X_3	X_4	X_5	X_6	X_7	X_8	X_9	X_{10}	X_{11}	X_{12}	X_{13}
X_1	0	0.52	−0.31	−0.96	−0.72	0.15	0.88	0.34	−0.18	−0.98	0.49	−1	−0.08
X_2	−0.47	0	−0.94	−0.14	0.11	0.66	0.52	0.3	−0.29	−0.69	−0.79	0.93	−0.22
X_3	0.85	0.58	0	−0.91	0.29	−1	−0.91	−0.94	0.93	−0.67	0.46	−0.3	0.31
X_4	0.87	0.38	−0.99	0	0.99	−0.94	−0.98	−1	0.45	−0.2	0.88	0.27	−0.99
X_5	−0.82	−0.11	−0.35	−0.3	0	−0.34	−0.27	0.83	0.92	0.22	−0.96	−0.88	0.27
X_6	−0.87	−0.74	−0.82	−0.49	0.96	0	0.06	0.74	0.99	−0.8	−0.21	−0.26	−0.15
X_7	−0.24	−0.25	0.02	−0.69	−0.69	−1	0	−0.98	−0.59	0.04	−0.46	0.42	−0.17
X_8	−1	−0.61	−0.25	−0.98	−0.35	−0.98	−0.59	0	0.82	0.58	0.31	−0.95	−0.23
X_9	−0.43	0.24	−0.86	−0.07	−0.5	−0.64	−1	0.12	0	0.18	−0.59	−0.85	0.4
X_{10}	−0.99	−0.15	0.13	−0.18	−0.15	−0.99	0.17	−0.35	−0.76	0	0.69	0.29	−0.89
X_{11}	0.49	−0.14	−0.24	−0.72	0.36	−0.98	−0.41	−1	−0.61	−0.38	0	0.59	−0.66
X_{12}	0.32	0.07	−0.97	0.21	−0.92	−0.13	−0.58	−0.59	−0.95	−0.22	−0.92	0	−0.42
X_{13}	−0.97	0.18	0.41	−1	0.79	−0.97	0.49	−0.85	0.94	−0.04	0.7	0.63	0

Table 9 Connection matrix for the FCM learned with the use of the approach based on density

$w_{j,i}$	X_1	X_2	X_3	X_4	X_5	X_6	X_7	X_8	X_9	X_{10}	X_{11}	X_{12}	X_{13}
X_1	0	0	−0.95	0	0	−0.64	0	−0.41	0	0	−0.98	−0.37	−0.35
X_2	0	0	0	−0.63	−0.97	−0.73	0	0	0	1	0.69	0	0
X_3	0	0	0	0	0	0	0	0	−0.99	0.63	0	0	−0.93
X_4	0	0.06	0	0	0.98	−0.88	0	0	−0.84	0.68	0	−0.63	0
X_5	0	0	0	0	0	0	−0.99	0	−0.68	−0.87	−0.71	0.98	−1
X_6	0	0	−0.88	−0.99	0.49	0	0	−0.67	0	0	−0.52	0	0
X_7	−0.84	0	0	0	−0.64	−0.9	0	0.27	−0.2	−0.36	0.07	−0.79	0
X_8	0.09	0	−1	0	0	0	0	0	−0.18	−0.6	0	0	0
X_9	0	−0.12	−0.24	0	0	−0.96	0	0	0	−0.46	0	0	−0.09
X_{10}	−0.78	0	−0.99	0	0	0	0	−0.67	0.95	0	−0.84	0	0
X_{11}	−0.99	0	−0.99	−0.99	0	−0.33	−0.42	−1	0	0	0	0	−0.99
X_{12}	−0.97	0	−0.62	−0.99	0	−0.43	0	0	0	−0.74	−0.94	0	0
X_{13}	0	0	−1	−0.96	0	−0.67	−0.96	−1	0	0	0	0	0

Table 10 Connection matrix for the FCM learned with the use of the proposed approach

$w_{j,i}$	X_1	X_2	X_3	X_4	X_5	X_6	X_7	X_8	X_9	X_{10}	X_{11}	X_{12}	X_{13}
X_1	0	0.54	0	−0.39	−1	0	0.32	−0.97	0	−0.95	0	−0.98	0.09
X_2	−0.38	0	0	0	0	0	−0.99	0	−0.1	0	0	0.35	−0.89
X_3	−1	0	0	−0.35	0	−0.99	0	−0.52	−0.94	0	0	0	0
X_4	0.51	0	0	0	0	0	−0.8	0	0	0.99	−0.96	−0.96	0.21
X_5	0.62	0	0	−1	0	−0.87	0.89	0	−0.32	0	0	−0.89	−0.6
X_6	0.22	0	0	−0.39	−0.64	0	0	−0.89	−1	0	0.96	0	0
X_7	−0.87	0	−0.89	−0.79	0	−0.79	0	−0.98	0	−0.53	−0.8	0.93	0
X_8	0	0.84	−0.71	0	−0.99	0	0	0	0.2	−0.64	0	0	0.56
X_9	−0.93	0	−1	−0.48	0	−0.73	−0.77	−0.09	0	0	0	0	0
X_{10}	0.38	0	0	0	0	−0.94	0	−0.17	0	0	−0.53	0	−0.99
X_{11}	0	−0.65	−0.98	−0.97	0	0	0	0	0	0	0	0.5	0.32
X_{12}	−0.92	0	−0.7	0	0	−0.93	−0.45	0	0.13	0.24	−0.35	0	0
X_{13}	−0.94	−0.56	−0.97	−0.91	0	0	0.32	−0.47	0	−0.99	−0.87	0	0

Table 11 System performance indicators obtained for the analyzed approaches

	X_1	X_2	X_3	X_4	X_5	X_6	X_7	X_8	X_9	X_{10}	X_{11}	X_{12}	X_{13}
STD $\overrightarrow{P_j}$	0.09	0.09	−0.24	−0.21	0.07	0.24	−0.24	0.2	0.24	−0.1	−0.22	0.09	0.2
STD $\overleftarrow{P_i}$	−0.22	−0.05	−0.07	−0.21	−0.08	0.07	−0.07	0.21	0.07	0.21	−0.07	0.22	0.21
DEN $\overrightarrow{P_j}$	0.15	−0.01	−0.07	−0.07	−0.22	−0.08	0.01	0.0	0.0	0.0	−0.16	−0.09	−0.23
DEN $\overleftarrow{P_i}$	−0.21	−0.03	−0.23	−0.09	−0.09	−0.29	−0.21	−0.08	0.15	−0.04	0.2	0.06	0.07
SPI $\overrightarrow{P_j}$	−0.35	−0.08	0.2	0.05	−0.07	0.2	0.07	−0.08	−0.35	0.34	−0.35	−0.33	−0.35
SPI $\overleftarrow{P_i}$	−0.07	−0.06	−0.08	0.07	−0.22	−0.22	−0.08	−0.07	0.06	0.22	−0.21	−0.22	−0.22

The obtained results show that the developed approach allows to approximate the historical data with satisfactory accuracy similar to the standard approach and the approach based on density. The best solutions for the analyzed approaches are presented below. Table 8 presents the resulting connection matrix obtained for the standard approach. Connection matrix for the approach based on density is shown in Table 9. Table 10 presents the connection matrix obtained for the proposed approach.

The FCM models obtained for the developed approach and method based on density are more readable and easier to interpret than the map obtained with the standard learning algorithm based only on data error. For further analysis, impact of the j-th concept on the system (9) and impact of the system on the i-th concept (10) were calculated and presented in Table 11. The developed approach allows to select only the most significant connections between concepts and receive the readable structure keeping satisfactory accuracy of modeling of historical data.

6   Conclusion

The paper presents the application of fuzzy cognitive maps with the developed evolutionary algorithm for fuzzy cognitive maps learning to model decision support systems based on real-life and historical data. The presented approach is based on data error, density and system performance indicators analysis. The obtained results show that the developed approach allows to approximate the available data with satisfactory accuracy similar to the standard approach and the approach based on density and receive the structure of the FCM model more similar to the fuzzy cognitive maps developed by experts. The resulting FCM models are readable and easier to interpret. The proposed approach allows to select only the most significant connections between concepts keeping satisfactory accuracy of modeling of data.

References

1. Acampora, G., Pedrycz, W., Vitiello, A.: A competent memetic algorithm for learning fuzzy cognitive maps. IEEE Trans. Fuzzy Syst. 23(6), 2397–2411 (2015)
2. Aguilar, J.: A survey about fuzzy cognitive maps papers. Int. J. Comput. Cogn. 3(2), 27–33 (2005)
3. Andreou, A.S., Mateou, N.H., Zombanakis, G.A.: Soft computing for crisis management and political decision making: the use of genetically evolved fuzzy cognitive maps. Soft Comput. J. 9(3), 194–210 (2005)
4. Borisov, V.V., Kruglov, V.V., Fedulov, A.C.: Fuzzy Models and Networks. Publishing house Telekom, Moscow (2004). (in Russian)
5. Buruzs, A., Hatwagner, M.F., Pozna, R.C., Koczy, L.T.: Advanced learning of fuzzy cognitive maps of waste management by bacterial algorithm. In: IFSA World Congress and NAFIPS Annual Meeting (IFSA/NAFIPS), pp. 890–895 (2013)
6. Chi, Y., Liu, J.: Learning of fuzzy cognitive maps with varying densities using multi-objective evolutionary algorithms. IEEE Trans. Fuzzy Syst. 24, 71-81 (2015)

7. De Vito, S., Massera, E., Piga, M., Martinotto, L., Di Francia, G.: On field calibration of an electronic nose for benzene estimation in an urban pollution monitoring scenario. Sens. Actuators B: Chem. **129**(2), 750–757 (2008)
8. Jastriebow, A., Poczeta, K.: Analysis of multi-step algorithms for cognitive maps learning. Bull. Pol. Acad. Sci. Tech. Sci. **62**(4), 735–741 (2014)
9. Kosko, B.: Fuzzy cognitive maps. Int. J. Man Mach. Stud. **24**(1), 65–75 (1986)
10. Kubuś, Ł.: Individually directional evolutionary algorithm for solving global optimization problems - comparative study. Int. J. Intell. Syst. Appl. (IJISA) **7**(9), 12–19 (2015)
11. Kubuś, Ł., Poczeta, K., Yastrebov, A.: A new learning approach for fuzzy cognitive maps based on system performance indicators. In: 2016 IEEE International Conference on Fuzzy Systems, Vancouver, Canada, 1398–1404 (2016)
12. Mateou, N.H., Andreou, A.S.: A framework for developing intelligent decision support systems using evolutionary fuzzy cognitive maps. J. Intell. Fuzzy Syst. **19**(2), 150–171 (2008)
13. Michalewicz, Z.: Genetic Algorithms + Data Structures = Evolution Programs. Springer, New York (1996)
14. Miettinem, K.M.: Nonlinear Multiobjective Optimization. Kluwer Academic Publishers, Boston (1999)
15. Poczeta, K., Yastrebov, A., Papageorgiou, E.I.: Learning fuzzy cognitive maps using structure optimization genetic algorithm. In: Ganzha, M., Maciaszek, L., Paprzycki, M. (eds.) Annals of Computer Science and Information Systems. Proceedings of the 2015 Federated Conference on Computer Science and Information Systems, vol. 5, pp. 547–554 (2015)
16. Poczeta, K., Kubuś, Ł., Yastrebov, A., Papageorgiou, E.I.: An economic decision support system based on fuzzy cognitive maps with evolutionary learning algorithm, position papers of the 2016 federated conference on computer science and information systems. In: Ganzha, M., Maciaszek, L., Paprzycki, M. (eds.) ACSIS, vol. 9, pp. 95–101 (2016)
17. Rodriguez-Repiso, L., Setchi, R., Salmeron, J.L.: Modelling IT projects success with fuzzy cognitive maps. Expert Syst. Appl. **32**, 543–559 (2007)
18. Salmeron, J.L.: Fuzzy cognitive maps for artificial emotions forecasting. Appl. Soft Comput. **12**, 3704–3710 (2012)
19. Silov, V.B.: Strategic Decision-Making in a Fuzzy Environment. INPRO-RES, Moscow (1995). (in Russian)
20. Słoń, G.: Application of models of relational fuzzy cognitive maps for prediction of work of complex systems. In: Lecture Notes in Artificial Intelligence LNAI, vol. 8467, pp. 307–318. Springer (2014)
21. Stach, W., Kurgan, L., Pedrycz, W., Reformat, M.: Genetic learning of fuzzy cognitive maps. Fuzzy Sets Syst. **153**(3), 371–401 (2005)
22. Stach, W., Pedrycz, W., Kurgan, L.A.: Learning of fuzzy cognitive maps using density estimate. IEEE Trans. Syst. Man Cybern. Part B **42**(3), 900–912 (2012)
23. Yesil, E., Urbas, L.: Big bang: big crunch learning method for fuzzy cognitive maps. World Acad. Sci. Eng. Technol. **71**, 815–8124 (2010)

Meeting the Challenges of Optimized Memory Management in Embedded Vision Systems Using Operations Research

Khadija Hadj Salem, Yann Kieffer and Stéphane Mancini

Abstract The ever growing complexity of signal and image processing applications, and the stringent constraints related to their implementation makes their design, simulation, and implementation more and more challenging. Memory management is among the main challenge that electronic designers have to face. In fact, it impacts heavily the main cost metrics, including area, performance (real-time aspect) and energy consumption, of modern-day electronic devices. For some particular cases of image treatments, with non-linear access patterns to the memory addresses, a co-designed architectural solution and its optimization process, called *Memory Management Optimization* (MMOpt), was proposed by Mancini et al. (Proc. DATE, 2012). It creates an ad-hoc memory hierarchy for accelerating the accesses to the memories holding large image data. This chapter studies the optimization challenge reflecting the efficient operation of the MMOpt tool, which is formalized as a 3-objective scheduling problem. New algorithms are proposed for producing efficient solutions, leading to enhance the run-time performance and reduce both energy consumption and cost of the circuits produced by MMOpt. The performance of these algorithms is compared, on the same real-world data set as used by Mancini et al. [14], against the one currently in use in the MMOpt tool. The results show that our algorithms perform well in terms of computational efficiency and solution quality.

K. Hadj Salem (✉) · Y. Kieffer
University of Grenoble Alpes, LCIS, 26000 Valence, France
e-mail: khadija.hadj-salem@lcis.grenoble-inp.fr

Y. Kieffer
e-mail: yann.kieffer@lcis.grenoble-inp.fr

S. Mancini
University of Grenoble Alpes, TIMA, 38031 Grenoble, France
e-mail: stephane.mancini@imag.fr

© Springer International Publishing AG 2018
S. Fidanova (ed.), *Recent Advances in Computational Optimization*,
Studies in Computational Intelligence 717, DOI 10.1007/978-3-319-59861-1_11

1 Introduction

The design of embedded vision systems carries many challenges, one of which is the efficient access to the image memory. This calls for new methods, tools, algorithms, and architectures that can help circuit designers meet their goals. An architectural solution, called *Memory Management Optimization* (MMOpt), was proposed by Mancini et al. [14] in the form of a software tool that creates an ad-hoc memory hierarchy for non-linear image accesses. But operating this kind of systems is itself an optimization challenge. We formalized this electronic problem as a scheduling problem, involving 3 objectives reflecting 3 main electronic design characteristics. They correspond to the energy consumption, performance, and size/cost of the circuit. To the best of our knowledge, this problem has not been studied before in the Operations Research (OR) literature.

In this chapter, we first give a brief description of the MMOpt design software arising in the context of embedded vision systems, and a clear explanation about the related optimization problematic set by the efficient operation of the circuits produced by MMOpt. A specific multi-objective mathematical model for this problem is then detailed and the basic assumptions are listed, as well as several sub-problems of interest. We then review the state of the art. After giving lower bounds for the 3-objective problem, we analyze the complexity of some of the mono-objective sub-problems. The description of new approaches is then given. Numerical experiments follow, which are conducted on real-world data for validating their efficiency, and a conclusion and perspectives section closes this chapter.

2 Embedded Vision Systems Context: Architectural Solution and Optimization Problematic

2.1 The Memory Management Optimization Tool

Among modern-day electronic devices, embedded vision systems such as picture and video cameras represent a specific design challenge with respect to memory management. Image sizes are measured in 100s of Kbs or even Mbs, while the access times must be short enough to allow the quick handling of the input image. For example, a live video feed may have 30 frames per second, meaning that the handling of one image (frame) must take less than 1/30 s. But it is a well-known fact in electronic memory design that access times grow with the size of the memory to be accessed. Due to this fact, it is not possible to reach good performance using only the standard memories. Something has to be added to improve the access times.

In image processing, a *kernel* denotes digital image treatment that works by operating each pixel value of the output image using straightforward mathematical operations to construct a new image. For the case of kernels that have linear access patterns to the memory addresses, usual caches as used for CPUs will solve this problem.

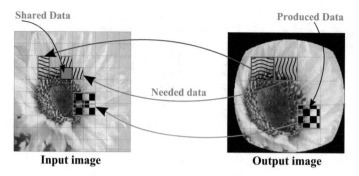

Fig. 1 Example of a non-linear kernel

However, the problem remains a big hurdle for the easy and efficient design of non-linear kernel circuit designs.

Figure 1 gives an example of a non-linear kernel, namely the *fisheye transform*. As shown in this figure, to compute a pixel $O(x, y)$ of the output image, the kernel needs a pixel $I(x', y')$ of the input image using a non-linear mapping function $(x, y) \mapsto (x', y')$. The pixels of the output image are computed by iterating over the coordinates (x, y). For each loop iteration the kernel makes a reference to the corresponding pixels (single pixel or a few adjacent pixels) in the input image.

The main difficulty of optimizing the memory management of non-linear kernels relies in the disparity of the function to compute the indices of the input data. An effective strategy to increase the data reuse and to optimize a memory hierarchy is to subdivide the input image into *tiles*, also defined as a set of pixels, that are small enough to fit the local memory and that can be processed independently. In this scheme, the output image is produced tile by tile, one after the other, while each output tile requires a fixed set of input tiles (needed data) from the input image, whatever the non-linear kernel. In Fig. 1, we can see that each set of required input tiles (left image) varies both in shape and in area, as well as in the case of output tiles (right image).

In order to deal with the non-linearity of kernels, a co-designed architectural solution, that creates ad-hoc memory hierarchies, was proposed by Mancini et al. [14] to address this challenge. This software tool, called *Memory Management Optimization* (MMOpt), takes as input a non-linear kernel for which the memory hierarchies is to be produced, such as the one shown in Fig. 1; it analyzes its access patterns; it then designs a run-time behavior for the whole resulting block, a so-called *Tile Processing Unit* (TPU); and it finally outputs the design of the TPU, together with the information needed to orchestrate its operational behavior.

We give some details about the architecture of the TPU, as shown in Fig. 2. It is made of (a) a *Prefetching Unit* (PU) that loads data from external memory to local buffers, and (b) a *Processing Engine* (PE), that implements the kernel and requests the prefetched data into input tiles buffers to compute output data. MMOpt computes and encodes into the TPU a schedule of prefetches, a mapping of tiles to buffers,

Fig. 2 Architecture
template of the TPU

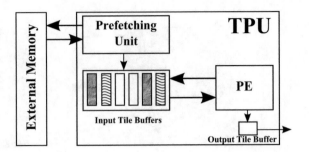

and a schedule of computations. Hence memory accesses in the final system are deterministic (i.e. independent of pixel values), and this is a requirement of the input kernel for the whole MMOpt scheme to work out.

2.2 The MMOpt's Optimization Challenge

When designing electronic circuits, some of the important design criteria are the area of the circuit produced, since it is directly related to production costs; the energy consumption, which may be limited, and which conditions the battery life for battery-powered devices; and the performance, which is usually a design parameter reflecting reactivity, and fluidity in the case of moving images.

TPUs produced by MMOpt embed schedules for the prefetches of input tiles and the computations of output tiles (see Fig. 3). In this figure, it is also possible to have pauses in between computations, so as to limit the number of necessary buffers.

The architecture of the TPU and those schedules will impact the three design characteristics in the following way: the number of buffers of the TPU will account for most of its area; the number of prefetches reflects the main part of the energy consumption; and the performance is related to the completion time of the whole prefetches-computations schedule for the computation of one image.

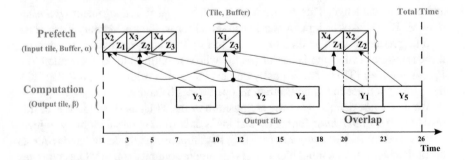

Fig. 3 Prefetches and computations schedules

Since MMOpt is a fully automatic electronic design software, computing good schedules is both a necessity and an opportunity for the circuit designers to deliver, with the help of MMOpt, low-cost, low-energy and efficient TPUs. Thus, this chapter introduces a set of new optimization algorithms that allow us to provide efficient solutions for MMOpt's user.

3 The 3-Objective Process Scheduling and Data Prefetching Problem

It is well known that the *Integer Linear Programming* (ILP) formulation is used in the OR as a modeling approach for the optimization problems. In this study, we have chosen to formulate our optimization issue by an off-line 3-objective scheduling model, with clearly delineated inputs and outputs, which we now present. This multi-objective formulation is a very flexible modeling approach that allows to deal precisely with specific sub-problems.

3.1 Problem Statement and Assumptions

The main multi-objective optimization problem considered in this paper is called *3-objective Process Scheduling and Data Prefetching Problem* (3-PSDPP). It involves the definition of the scheduling of output tiles computations and the scheduling of input tiles prefetches, while meeting a requirement constraint between prefetches and computations, and simultaneously minimizing the number of prefetches, the number of buffers of the TPU, and the total completion time.

Before giving a formal presentation of 3-PSDPP, we list some assumptions to clarify some constraints that all TPUs produced by MMOpt, which are set up by Mancini et al. [14], have to satisfy. These assumptions can be summarized as follows:

- Input tile sizes are identical and each input tile fits exactly into one buffer.
- There is no distinction between buffers, i.e. any input tile may be prefetched into any buffer.
- All input (respectively output) tiles and the subset of input tiles required to compute each output tile are known in advance.
- Only one input (output) tile can be prefetched (computed) at a time.
- The prefetch operations and the computation steps may be carried out simultaneously.
- Input (output) tile prefetch (respectively computation) times are constant and identical.

Table 1 Mathematical formulation for 3-PSDPP

Inputs	$\mathcal{X} = \{1, \ldots, X\}$, $\mathcal{Y} = \{1, \ldots, Y\}$, where $X, Y \in \mathbb{N}^*$
	$\mathcal{R}_y \subseteq \mathcal{X}, \forall\, y \in \mathcal{Y}$
	$\alpha, \beta \in \mathbb{N}^*$
Outputs	$(p_i)_{i \in \mathcal{N}}$, where $p_i = (d_i, b_i, t_i)$ and $\mathcal{N} = \{1, \ldots, N\}$
	$(c_j)_{j \in \mathcal{M}}$, where $c_j = (s_j, u_j)$, $\mathcal{M} = \{1, \ldots, M\}$ and $M = Y$
	(N, Z, Δ), where $N, Z, \Delta \in \mathbb{N}^*$
Constraints	**(1)** $\forall\, y \in \mathcal{Y}, \exists\, j \in \mathcal{M} \;/\; s_j = y$
	(2) $\forall\, j \in \mathcal{M}, \forall x \in \mathcal{X}, x \in \mathcal{R}_{s_j} \Rightarrow (\exists\, a \in \{1, \ldots, u_j - \alpha\}, \exists\, i \in \mathcal{N}/t_i = a, d_i = x \;\&\; (\forall a' \in \{a + \alpha, \ldots, u_j + \beta - 1\}, \forall i' \in \{i + 1, \ldots, N\}, t_{i'} = a' \Rightarrow b_{i'} \neq b_i))$
	(3) $\forall\, i \in \mathcal{N} \backslash \{1\}, t_i \geq t_{i-1} + \alpha$
	(4) $\forall\, j \in \mathcal{M} \backslash \{1\}, u_j \geq u_{j-1} + \beta$
Objectives	$\min N, \min Z, \min \Delta$

3.2 Formulation for 3-PSDPP

We now describe the input data of our 3-PSDPP scheduling problem, the expected output, the constraints, and the 3 formal objectives reflecting the 3 electronic design characteristics. A mathematical formulation is summarized in Table 1.

3.2.1 Inputs

A 3-PSDPP instance is represented by a 5-tuple $(\mathcal{X}, \mathcal{Y}, (\mathcal{R}_y)_{y \in \mathcal{Y}}, \alpha, \beta)$, where $\mathcal{X}$ is the set of input tiles to be prefetched, and $\mathcal{Y}$ is the set of output tiles to be computed successively without preemption. Each output tile y requires its own set of input tiles, denoted by $\mathcal{R}_y$. Also, the duration of a prefetch step α, and that of a computation step β, have to be given as input.

3.2.2 Outputs

A feasible solution to such an instance is defined by $((p_i)_{i \in \mathcal{N}}, (c_j)_{j \in \mathcal{M}}, Z, N, \Delta)$, which are described as follows:

- **Configuration of the prefetched input tiles**: we denote by $(p_i)_{i \in \mathcal{N}}$ the prefetch sequence, where $p_i = (d_i, b_i, t_i)$ encodes which input tile d_i is prefetched in which buffer b_i and at which time t_i.
- **Configuration of the computed output tiles**: we denote by $(c_j)_{j \in \mathcal{M}}$ the computation sequence, where $c_j = (s_j, u_j)$ encodes which output tile s_j is to be computed at which time u_j.

- **The values for the three criteria** (Z, N, Δ): we denote by Z the number of buffers; N is the total number of prefetched input tiles; and Δ is the total completion time, meaning the total time it takes for the whole operation of the TPU from the beginning of the first prefetch to the end of the last computation of one full image.

3.2.3 Constraints

The first constraint (1) on solutions is that for each output tile y, there exists a computation step j in which this output tile is computed. The second and main constraint (2) ensures that all the input tiles $\mathcal{R}_y$ required by y have to be prefetched from the external memory to the internal storage area (buffers) before the start date u_j of its associated computation step, and will not be overwritten until its end date. Input tiles already prefetched earlier can be reused, provided they have not been overwritten. Constraints (3) and (4) guarantee that different input (output) tiles cannot be prefetched (computed) simultaneously.

3.2.4 Objectives

In the formulation above, the three objectives have to be minimized are: (i) the number of prefetches N reflects the main part of the energy consumption due to the data transfer between external memory and the TPU; (ii) the number of buffers Z of the TPU will account for most of its area, and is related to cost; and (iii) the completion time Δ accounts for the performance of the TPU.

3.3 Sub-problems of 3-PSDPP

From this multi-objective problem 3-PSDPP, we derive several mono- and bi-objective sub-problems. The three natural mono-objective sub-problems are:

- **Minimum Buffers of 3-PSDPP (MB-PSDPP)**, in which the number of buffers Z is to be minimized.
- **Minimum Prefetches of 3-PSDPP (MP-PSDPP)**, in which the number of prefetches N is to be minimized.
- **Minimum Completion Time of 3-PSDPP (MCT-PSDPP)**, in which the completion time Δ is to be minimized.

For the remaining sub-problems to be presented, the number of buffers Z will be fixed as input data. Hence, we consider three variants which are described as follows:

- **Prefetching and Scheduling Problem (PSP)**, where the number of buffers Z is fixed as input, the completion time Δ has to be determined as an output data and the number of prefetches N is to be minimized.

- **Data Prefetching Problem (DPP)**, which is the variant of PSP, where the computation sequence is given as part of the input.
- **Buffer-Constrained Minimum Completion Time Problem (B-C-MCTP)**, where the number of buffers Z is fixed as input, the number of prefetches N has to be determined as an output data and the completion time Δ is to be minimized.
- **2-objective Process Scheduling and Data Prefetching Problem (2-PSDPP)**, where the number of buffers Z is fixed, and both N and Δ are to be minimized.

4   State of the Art

We now review the state of the art regarding the previous work that proposed by Mancini et al. [14, 15] and the relation with some closest classical optimization problems found in the OR literature.

4.1   Previous Work on MMOpt's Optimization

Since the use of OR methods for optimizing the running of the TPU produced by the MMOpt tool is still an emerging field, we found only one systematic study of the published literature of MMOpt from 2012, done by Mancini et al. [14, 15]. This study is the only generic proposition that allows a significant performance improvement, and that is applicable to any non-linear kernel. It only gives a description of two algorithms that provide solutions, without defining a formal sketch, to the related optimization problem which represents the main focus of our work.

The first algorithm is M_1, for which both the number of prefetches N and the completion time Δ are minimized. The second one, called M_2, aims at minimizing both the number of prefetches N and the number of buffers Z. The algorithm M_1 proceeds in three steps which we call, respectively, *Computations*, *Prefetches*, and *Destinations*. Furthermore, M_2 comprises four steps, the first two, and the last of which are those of M_1. The third step is *Delay Computations*.

We now give some explanations about the flowchart shown in Fig. 4, which outlines the various steps already mentioned in the previous paragraph.

- **Step1 - Computations**: this step encodes the order in which a batch of output tiles has to be successively computed, one at a time. The traffic to the external memory is minimized by performing this step.

  To construct the computation sequence, Mancini et al. [15] solve an instance of the *Asymmetric Traveling Salesman Problem* (ATSP) to find a *Hamiltonian Path* in the complete directed graph $\overrightarrow{G}$ whose vertices are the set of output tiles, and whose arcs are weighted by $\varphi(k, l)$, where (k, l) is a pair of output tiles, in which the output tile k will be computed before the output tile l. The function $\varphi(k, l)$ defines the number of additional input tiles to be prefetched for computing the

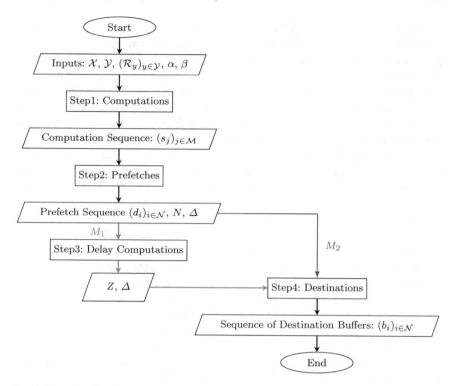

Fig. 4 Flowchart for algorithms M_1 and M_2

output tile l ($|\mathcal{R}_l \backslash \mathcal{R}_k|$), when all input tiles shared between k and l are already prefetched.

- **Step2 - Prefetches**: in this step, the authors determine the schedule of prefetches associated to the computation sequence given by step 1. This schedule encodes which input tile should be prefetched from the external memory to the buffers at each moment. In fact, in parallel to each computation step, they prefetch the additional input tiles needed for the next computation.
- **Step3 - Delay Computations**: in this step, in order to reduce buffer usage, the authors simply delay some computations when necessary.
- **Step4 - Destinations**: this step computes in which buffer to place each prefetched input tile.

4.2 Related Problems from the OR Literature

To the best of our knowledge, the 3-PSDPP scheduling problem introduced in this paper has not been studied before in the OR literature. We now relate some of its sub-problems to similar ones in the OR literature.

We first focus on the uniform variant of *Tool Switching Problem* (ToSP) arising in the flexible manufacturing context. In this problem, we consider a set of jobs to be processed sequentially without preemption on a single machine, and a set of tools to be loaded on a magazine with a limited capacity. Each job requires a subset of tools to be loaded on the tool magazine of the machine before the job can be processed. In most practical situations, the magazine cannot hold all tools at once, so that some tool switches may be necessary when performing two jobs in succession. A *tool switch* consists of removing a tool from the magazine and inserting another one in its place.

The ToSP involves optimally sequencing jobs and assigning tools to a capacitated magazine in order to minimize the total number of tool switches. Several variants and extensions of the ToSP are surveyed in Blazewicz and Finke [7], in Crama [9] and in Balakrishnan and Chakravarty [3].

The general ToSP was first considered by Tang and Denardo [17]. They showed that the ToSP for a given job sequence, called *Tooling Problem* (TP), can be solved in polynomial time by means of a *Keep Tools Needed Soonest* (KTNS) algorithm. The KTNS policy states that when tool changes are necessary, the tools required the soonest for an upcoming job should be kept first in the magazine.

On the other hand, when the job sequence is to be determined, Crama et al. [10] showed that the ToSP is already NP-Hard for any fixed tool magazine capacity larger than or equal to 2. Different optimization techniques, including exact — based on *Integer Linear Programming* (ILP) formulation — and heuristic methods, have been applied to its resolution (see Bard [4]; Privault et Finke [16]; Laporte et al. [13]; Konak et al. [12]; Amaya et al. [1]; Catanzaro et al. [8]).

5 Models Analysis

For validating the efficiency of the proposed approaches, we first develop three lower bounds lb_N, lb_Z, and lb_Δ for the different optimization criteria (N, Z, Δ). We then give complexity results for the different mono-objective sub-problems of 3-PSDPP problem described in Sect. 3.3.

5.1 Lower Bounds

Proposition 1 $X - |\Omega|$ *is a lower bound on the number of prefetches for the 3-PSDPP, where Ω denotes the set of input tiles which are not required by any output tile.*

Proof For any solution to some given instance of 3-PSDPP, all input tiles that are required at least once for the computation of an output tile have to be prefetched at least once to some buffer. Hence the total number of prefetches cannot be less than $X - |\Omega|$.

Proposition 2 $\max_{y \in \mathcal{Y}} |\mathcal{R}_y|$ *is a lower bound on the number of buffers for the 3-PSDPP.*

Proof Fix an instance of 3-PSDPP, and a feasible solution for that instance. When an output tile is computed, all the required input tiles have to be simultaneously present in the buffers. Hence $\max_{y \in \mathcal{Y}} |\mathcal{R}_y|$ is a lower bound for the number of buffers in the solution.

Proposition 3 $lb_1 = \alpha * X' + \beta$ *and* $lb_2 = \alpha + \beta * Y$ *are lower bounds on the completion time* Δ *for the 3-PSDPP.*

Proof Fix an instance of 3-PSDPP, and a feasible solution for that instance. Since all the input tiles have to be loaded before the last computation starts, the completion time is at least $\alpha * X'$ (for the prefetches) plus β (for the computation of the last output tile).

Likewise, all output tiles have to be computed, and no computation can start before a first input tile has been prefetched. Hence the completion time is lower bounded by $\beta * Y$ (computation time for all output tiles) plus α (prefetch time for the first prefetch).

Thus, the completion time Δ is lower bounded by the maximum between the two lower bounds lb_1 and lb_2: $lb_\Delta = \max\{lb_1, lb_2\}$.

5.2   Basic Complexity Results

Theorem 1   *MB-PSDPP is solvable in polynomial time (MB-PSDPP $\in$ P)*

Proof Consider a particular instance of MB-PSDPP, we give a polynomial algorithm for which the number of buffers Z equals its lower bound ($Z_{min} = lb_Z$).

The main idea of this algorithm is that the prefetch steps can not be carried out in parallel with the computation steps. In this method, we first fix the number of buffers Z to $\max_{y \in \mathcal{Y}} |\mathcal{R}_y|$. Then, for each output tile, we prefetch all its required input tiles into the Z buffers before the corresponding computation step starts.

Theorem 2   *MP-PSDPP is solvable in polynomial time (MP-PSDPP $\in$ P)*

Proof To any given instance of MP-PSDPP, we give a polynomial algorithm for which the number of prefetches N equals its lower bound ($N_{min} = lb_N$).

In this method, we first prefetch successively all the input tiles in $\mathcal{X}'$, where $\mathcal{X}' = \mathcal{X} \backslash \Omega$ and Ω denotes the set of the input tiles which are not required by any output tile. Then, when the prefetch steps are finished, all output tiles are successively computed.

Theorem 3   *PSP is NP-hard for any fixed number of buffers $Z \geq 2$*

Theorem 4 *DPP is solvable in polynomial time (DPP ∈ P)*

Proof Consider a particular instance of PSP, we have proved the equivalence between PSP and the ToSP, which is described in Sect. 4.2. In the description of the PSP problem, both input and output tiles $(\mathcal{X}, \mathcal{Y})$ are regarded as ToSP data (tools, jobs). The incidence matrix Tools×Jobs can then be regarded as the requirements of input tiles needed to compute all the output tiles $(\mathcal{R}_y)_{y \in \mathcal{Y}}$. The fixed number of buffers Z is the analogue of the capacity of the tool magazine. In addition, finding a computation sequence for minimizing the total number of prefetches corresponds to finding a job sequence for minimizing the total number of tools loadings. Thus, PSP is NP-Complete.

When the the computation sequence is given as input data, we have proved the equivalence between DPP and the TP, a polynomial variant of ToSP. Hence, DPP is polynomially solvable.

This equivalence allows us to adapt the KTNS (Keep Tools Needed Soonest) algorithm, as described by Tang and Denardo [17] for solving the TP, to give an optimal solution for DPP. We call this adaptation *KTNS (Keep Tiles Needed Soonest) Adapted to DPP* (KAD). A more detailed description of this specific adaptation can be found in [11]. On the other hand, in the case of PSP, we have developed an algorithm, named *KTNS Adapted to PSP* (KAP), to solve it.

Theorem 5 *B-C-MCTP is NP-Complete*

Proof Note that the B-C-MCTP can be easily solved by a polynomial time nondeterministic algorithm (B-C-MCTP ∈ NP), we then give a polynomial-time reduction from the "Edge Hamiltonian Path" (EHP) problem to B-C-MCTP problem: EHP $\propto$ B-C-MCTP.

The EHP problem is a combinatorial optimization problem for general graphs. Let $G = (V, E)$ be an undirected graph comprising a set V of vertices together with a set E of edges and $H = (E, I)$ be its edge-graph, where each vertex E of H is an edge of G and each pair $\{e, f\}$ $(e, f \in E)$ is an edge of H if and only if e and f share a common vertex in G. The EHP problem consists in finding a *Hamiltonian Path* in H. Note that a *Hamiltonian Path* is a path that visits each vertex exactly once.

We are going to prove that the EHP problem can be formulated as a special case of the B-C-MCTP problem, where Z is fixed as input $(Z = \max_{y \in \mathcal{Y}} |\mathcal{R}_y|)$, $\beta = 1$ and $\alpha = \beta * Y = Y$ units of time. Let $G = (V, E)$, where $V = \{1, 2, \ldots, X'\}$ and $E = \{e_1, e_2, \ldots, e_{Y'}\}$, be an instance of the EHP problem, we then define an $X \times Y$ matrix $\mathcal{R}$, with rows associated to X' vertices of G, columns associated to the Y' edges of G, and such that $r_{xy} = 1$, if edge e_y contains vertex x, and 0 otherwise. Note that the number of prefetches N is the only relevant information to the B-C-MCTP under consideration. Meaning that the completion time can be determined, in this case, by the quantity $\alpha * N + \beta = \alpha * N + 1$ (since $\beta = 1$).

Consider now $\mathcal{R}$ as an instance of the B-C-MCTP, where $Z = \max_{y \in \mathcal{Y}} |\mathcal{R}_y|$. A computation sequence for this problem corresponds to a permutation of E. Also, we can see that the number of prefetches between two successive computations j and k, corresponding to the edges e_j and e_k of G, equals $Z - 1$, and hence N equals $Y' + Z - 1$.

Therefore, finding an optimal computation sequence, with a minimum completion time Δ, amounts to finding a *Hamiltonian Path* in H, in which Δ equals $\alpha(Y' + Z - 1) + 1$. Since EHP is NP-Complete [6], we conclude that the general B-C-MCTP is NP-Complete.

For the MCT-PSDPP sub-problem, in which the completion time Δ is to be minimized, we have not yet been able to determine its complexity. In contrast, we have developed a set of three heuristics methods — called *Earliest Computations for MCT* (ECM), *Computation Grouping for MCT* (CGM) and *Computation Classes for MCT* (CCM) — to solve it. A detailed description of these algorithms will be given in Sect. 6.3.

6 Solution Methods

Dealing with a multi-objective optimization problem (3-PSDPP), with contradictory objectives such as low cost, low energy and high performance, it is difficult to find a unique solution to solve it. Therefore, an important question is the existence of polynomial time algorithms to find good solutions to it. We then focus on the development of polynomial constructive heuristics to solve some of its sub-problems. These approaches allow us to provide useful solutions for MMOpt's user.

6.1 Algorithm KAP for PSP

We have developed the *KTNS Adapted to PSP* (KAP) algorithm for solving the PSP sub-problem of 3-PSDPP, in which the number of buffers Z is fixed as input ($Z \geq \max_{y \in \mathcal{Y}} |\mathcal{R}_y|$), the total completion time Δ has to be determined and the number of prefetches N is to be minimized. In contrast to DPP, the order in which the computations have to be carried out is not given as input, and has to be determined.

The KAP algorithm proceeds in two steps as follows:

- **Step1 - Find a Computation Sequence**: in this step, we determine the computation sequence by solving the same instance of an ATSP as step 1 of both algorithms M_1 and M_2 (see Sect. 4.1).
- **Step2 - KAD Algorithm**: in this step, we determine the schedules of both the prefetches and computations with an optimal number of prefetches N by applying the polynomial *KTNS (Keep Tiles Needed Soonest) Adapted to DPP* (KAD) algorithm (see the proof of Theorem 4 given in Sect. 5.2).

Note that the KAD algorithm will be also an intermediate step in the SPbP algorithm presented in the following subsection.

6.2 Algorithm SPbP for 2-PSDPP

We have developed a solution approach called *Shifted Prefetches for bi-PSDPP* (SPbP) for solving the 2-PSDPP sub-problem of 3-PSDPP, in which the number of buffers Z is fixed as input ($Z \geq \max_{y \in \mathcal{Y}} |\mathcal{R}_y|$), and both the number of prefetches N and the completion time Δ are to be simultaneously minimized.

The SPbP algorithm proceeds in the following steps:

- **Step1 - Find a Computation Sequence**: this step is the same as the first step of the KAP algorithm described in Sect. 6.1.
- **Step2 - KAD Algorithm**: this step is the same as the second step of the KAP algorithm described in Sect. 6.1.
- **Step3 - "Shifting Prefetches"**: in this step, in order to minimize the total completion time Δ, we shift in the resulting schedules of prefetches — given by the KAP algorithm — those that can be carried out in parallel with the previous computation step, by checking that no required input tiles of the output tile in this step were overwritten until its end date. The new start dates of both the prefetches and computations schedules together with the value of the total completion time Δ are then computed.

6.3 Algorithms for MCT-PSDPP: ECM, CGM and CCM

For solving the MCT-PSDPP sub-problem of 3-PSDPP, in which the total completion time Δ is to be minimized, we have developed a set of three algorithms to solve it. They are called *Earliest Computations for MCT* (ECM), *Computation Grouping for MCT* (CGM) and *Computation Classes for MCT* (CCM). For all these algorithms, the number of prefetches N equals its lower bound lb_N and the number of buffers Z equals its number of required input tiles $\mathcal{X}'$, where $\mathcal{X}' = \mathcal{X} \backslash \Omega$.

We now give a brief description of the different steps for each one of them in the following subsections.

6.3.1 Algorithm ECM for MCT-PSDPP

The main idea of the ECM algorithm is to compute the output tiles at the earliest, while respecting the input tiles requirement constraint.

The ECM algorithm consists of two steps as follows:

- **Step1 - Find a Prefetches Schedule**: we first calculate the number of occurrences $O_c(x)$, for each input tile x in $(\mathcal{R}_y)_{y \in \mathcal{Y}}$. Then, the prefetches d_i, $i \in \mathcal{N}$ are sequenced in their decreasing order of $O_c(x)$, $\forall x \in \mathcal{X}$.
- **Step2 - Find a Computations Schedule "At Earliest"**: for each computation j, $j \in \mathcal{M}$, it is determined when it can be scheduled at the earliest. The corresponding date is the end of the loading of the latest prefeteched tile among its required input tiles. Computations s_j, $j \in \mathcal{M}$ are scheduled greedily in this order, while making sure to respect these "at earliest" dates.

6.3.2 Algorithm CGM for MCT-PSDPP

The main idea of the CGM algorithm is to find a set of groups $\mathcal{G}$, where a *Group $\mathcal{G}$* defines a set of output tiles y, $y \in \mathcal{Y}$ which share the same required input tiles. More formally, consider an output tile y, $y \in \mathcal{Y}$, a *Group $\mathcal{G}$* of y is defined by $\mathcal{G}(y) = \{g : g \in \mathcal{Y}, g \neq y, \text{ and } \mathcal{R}_g \subseteq \mathcal{R}_y\}$.

The CGM algorithm proceeds in two steps as follows:

- **Step1 - Find a Computation Sequence Using** *Groups G*: we first determine the set of groups $\mathcal{Y}'$, associated to the set output tiles y, $y \in \mathcal{Y}$, while ensuring that each output tile y, $y \in \mathcal{Y}$ belongs to exactly one group $\mathcal{G}$. Note that a *Group $\mathcal{G}(y)$* is a set of output tiles to be successively computed after y while the required input tiles by y are prefetched in the internal buffers. Then, the computations s_j, $j \in \mathcal{M}$ are sequenced in their increasing order of $|\mathcal{R}_y|$, $\forall y \in \mathcal{Y}'$ and $\mathcal{Y}' \subseteq \mathcal{Y}$.
- **Step2 - Find Prefetches and Computations Schedules**: in this step, for each output tile y, $y \in \mathcal{Y}'$, when the corresponding computation step j is started, we prefetch the set of input tiles which are required by the computation step $j + 1$ but have never been prefetched during all the previous computations $j - k$, $k \in \{1, \ldots, j-1\}$ and $j \in \{1, \ldots, |\mathcal{Y}'|\}$. This ensures that each input tile x, $x \in \mathcal{X}$ is prefetched only once. Then, prefetches d_i, $i \in \mathcal{N}$ are scheduled in this order, while making sure to respect the corresponding buffer, which means that each prefetch is performed in its own buffer. The start dates of both the prefetches and computations schedules together with the value of the total completion time Δ are then computed.

6.3.3 Algorithm CCM for MCT-PSDPP

The main idea of the CCM algorithm is to construct a set of classes $\mathcal{C}$, where a *Class C* defines a set of output tiles y, $y \in \mathcal{Y}$ that together require at most Z_0 input tiles. More formally, C is a *Class* if and only if $\left| \bigcup_{y \in C} \mathcal{R}_y \right| \leq \max_{y \in \mathcal{Y}} |\mathcal{R}_y|$.

The CCM algorithm proceeds in the following steps:

- **Step1 - Find a Computation Sequence Using** *Classes* C: this step consists in finding a feasible partition of classes $\mathcal{P}$, associated to the set of output tiles $\mathcal{Y}$. We make sure that each output tile y, $y \in \mathcal{Y}$ belongs to exactly one set P_k in $\mathcal{P}$ and P_k is a *Class*. Note that a *Class* C is a set of output tiles that can be computed without incurring any new input tile prefetches, while making sure that all its required input tiles are being available in the internal buffers.

  For the sequencing of the set of classes in this partition, we have proposed three ideas which can be outlined as follows:

  1. **CCM1**: we first calculate the number of shared input tiles $N_s(C)$, for each class C in $\mathcal{P}$. $N_s(C)$ defines the total number of shared input tiles between the class C and each class C' in $\mathcal{P} \backslash C$. More formally, given a pair of two classes (C, C'), the number $N_s(C)$ between C et C' equals $|C \cap C'|$: $N_s(C) = \sum_{C' \in \mathcal{P}} |C \cap C'|$. Then, the set of classes in $\mathcal{P}$ are sequenced in their decreasing order of $N_s(C)$, $\forall C \in \mathcal{P}$.

  2. **CCM2**: we first calculate the number of occurrence $N_o(C)$, for each class C in $\mathcal{P}$. $N_o(C)$ defines the total number of $O_c(x)$ for the class C, where $x \in \mathcal{R}_y(C) = \bigcup_{y \in C} \mathcal{R}_y$ and $O_c(x)$ represents the number of occurrences for the input tile x in $\mathcal{R}_y(C)_{C \in \mathcal{P}}$: $N_o(C) = \sum_{x \in \mathcal{R}_y(C)} O_c(x)$. Then, the set of classes in $\mathcal{P}$ are sequenced in their decreasing order of $N_o(C)$, $\forall C \in \mathcal{P}$.

  3. **CCM3**: in this case, the set of classes in $\mathcal{P}$ are sequenced in the decreasing order of the number of output tiles that belong to each class C: $|C|$, $\forall C \in \mathcal{P}$.

  Note that the set of output tiles in the same class C, in the context of the three ideas given above (CCM1, CCM2 and CCM3), are sequenced in their decreasing order of $|\mathcal{R}_y|$, $\forall y \in C$ and $\forall C \in \mathcal{P}$.

- **Step2 - Find Prefetches and Computations Schedules**: in this step, we compute successively the set of output tiles that belong to the same class $\mathcal{C}$, while making sure that their Z corresponding required input tiles $(|\bigcup_{y \in C} \mathcal{R}_y|)$ are prefetched and being available in the internal buffers. Then, prefetches are scheduled in this order, while making sure that each input tile x, $x \in \mathcal{X}$ is prefetched only once and each prefetch is performed in its own buffer. The start dates of both the prefetches and computations schedules together with the value of the total completion time Δ are then computed.

In the next section, we give an extension of the different heuristics under consideration by taking into account, the minimization of the different optimization criteria (N, Z and Δ) simultaneously.

6.4 Solutions for 3-PSDPP

As it has been mentioned in Sect. 2.2, our main optimization challenge related to the MMOpt tool is formalized as a 3-objective optimization problem, with contradictory objectives such as low cost (Z), low energy (N) and high performance (Δ). To solve it, we have developed four extensions of some of our proposed heuristics named, respectively, *Extended SPbP* (E-SPbP), *Extended ECM* (E-ECM), *Extended CGM* (E-CGM) and *Extended CCM* (E-CCM) for 3-PSDPP. The main idea of each one of them can be summarized as follows:

- **Algorithm E-SPbP for 3-PSDPP**: on the basis of the previously described SPbP algorithm (see Sect. 6.2), we can reuse it as an efficient resolution method for producing solutions to the main 3-PSDPP problem. By varying the number of buffers Z, from its lower bound lb_Z to an upper value uv_Z (may be defined as $uv_Z = \sum_{y \in \mathcal{Y}} |\mathcal{R}_y|$ or $uv_Z = X'$), MMOpt's user can explore the design space to choose his favorite compromise solution.
- **Extended MCT's Algorithms for 3-PSDPP**: for each one of the three algorithms (ECM, CGM and CCM), which were proposed for solving the MCT-PSDPP (see Sect. 6.3), we consider three extended versions to solve the main 3-PSDPP. These algorithms are E-ECM, E-CGM and E-CCM. For each one of them, we added a third step, called *Reduce Buffers Usage*, in order to minimize the number of buffers Z. In this step, we simply reuse some buffers when the corresponding prefetched input tile will no longer be used after its last associated computation step.

6.5 Flowcharts for Our Proposed Methods

We now give a set of four flowcharts — shown in Figs. 5, 6, 7 and 8 — that defines the list of shared or combined steps, between all our proposed algorithms.

The different steps of both KAP and SPbP algorithms are outlined in Fig. 5. As shown in this figure, the KAP algorithm represents the first two steps of the SPbP algorithm.

In the same way, the different steps of both ECM and E-ECM algorithms are outlined in Fig. 6, as well as the steps of both CGM and E-CGM algorithms which are outlined in Fig. 7. As shown in these figures, the ECM algorithm represents the first two steps of the E-ECM algorithm and, similarly, the CGM algorithm represents the first two steps of the E-CGM algorithm.

The flowchart of Fig. 8 summarizes the main steps of the CCM algorithm. As shown in this figure, the CCM algorithm represents the first two steps of the E-CCM algorithm.

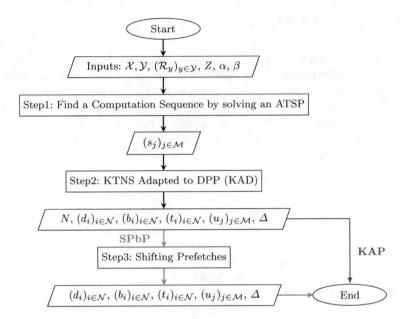

Fig. 5 Flowchart of KAP and SPbP algorithms

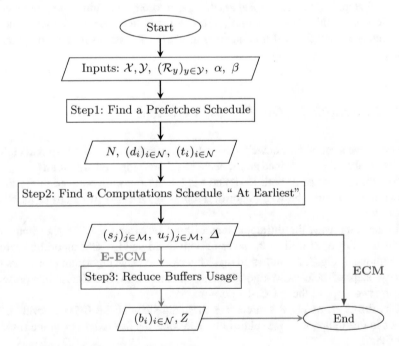

Fig. 6 Flowchart of ECM and E-ECM algorithms

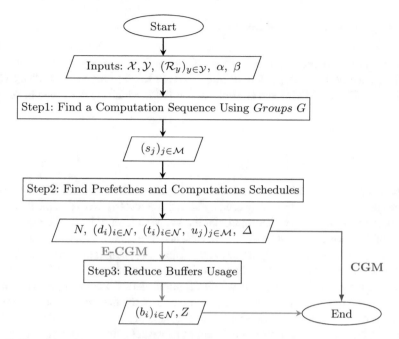

Fig. 7 Flowchart of CGM and E-CGM algorithms

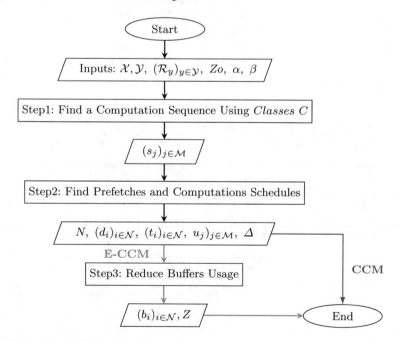

Fig. 8 Flowchart of CCM and E-CCM algorithms

7 Experiments and Results

In this section, we present a series of simple numerical experiments to evaluate the performance of the developed algorithms, namely KAP, SPbP, ECM, CGM, CCM, E-ECM, E-CGM, E-CCM and E-SPbP for 3-PSDPP, which are described in Sect. 6.

7.1 Data Sets

Experiments were conducted using a set of 5 benchmarks from real-life non-linear image processing kernels already used by Mancini et al. [14]. Note that the incidence matrices of the kernels are our input, not the image processed by the kernel.

Table 2 shows the characteristics of the test instances that were used, together with the values for X (number of input tiles) and Y (number of output tiles to be computed). As summarized in this table, the benchmarks are variations of four kernels (fisheye, polar, fd resize, and fd haar) for which the input data structure (multi-resolution (an)isotropic mipmap input data) is modified. In fact, the first four kernels represent geometric non-linear transformations (see Thornton et al. [18] and Bellas et al. [5]). The fifth kernel, which represents a kernel of a face detection (fd) application based on haar features, creates a pyramidal multi-resolution image (see Viola et al. [19]). The input image tiles number varies between 350 and 7000, and the number of the output tiles varies between 150 and 1200 tiles.

Table 2 also gives for each kernel the values of the different lower bounds $(lb_Z, lb_N, lb_1, lb_2, lb_\Delta)$, that are developed in Sect. 5.1. These lower bounds allow us to evaluate the performances of our proposed approaches.

Table 2 Parameter values of data sets and lower bounds for N, Z, and Δ

N°	Kernel	Input data type	X	Y	lb_Z	lb_N	lb_1	lb_2	lb_Δ
1	Fisheye	Mipmap isotropic	352	158	13	224	452	477	477
2	Fisheye	Mipmap anisotropic	704	158	21	360	724	477	724
3	Polar	Mipmap anisotropic	4225	112	20	244	492	339	492
4	Fd Resize	Mipmap isotropic	1280	1186	13	429	862	3561	3561
5	Fd Haar	Pyramidal integral image	7040	428	96	2272	4548	1287	4548

7.2 Numerical Results

This section presents an experimental analysis of the performance of the different algorithms. All the algorithms were coded in Python, except the ATSP part which was re-encoded as a TSP, and run through Concorde's implementation of the Chained Lin-Kernighan heuristic for the TSP (see Applegate et al. [2]). Tests were run on a computer powered by an Intel Core i5 processor clocked at 2.60 GHz. All our tests were carried out for the case where $\alpha = 2$ and $\beta = 3$ time units. The running time of all the different algorithms is in the order of a few minutes, which is very reasonable given the application context.

We first analyze the performance of both KAP and SPbP algorithms. We recall that the number of buffers Z is considered as an input data, where $Z \geq \max_{y \in \mathcal{Y}} |\mathcal{R}_y|$. In order to compare our proposed methods to the ones formerly used in the MMOpt tool (M_1 and M_2), we define two different values of Z, Z_1 and Z_2, as follows:

- The value of Z_1 is larger than the lower bound lb_Z ($Z_1 \gg lb_Z$), for which the MMOpt's algorithm M_1 reaches the maximum number of buffers and the completion time is minimized.
- Whereas, Z_2 gives the minimum number of buffers (Z_2 equals its lower bound lb_Z).

Table 3 summarizes the numerical results for both KAP and SPbP algorithms, where Z is fixed to Z_1, respectively to Z_2, and those of the algorithms M_1, and M_2 on different data sets described in Table 2. For the 5 kernels shown on row 1, the third row gives the number of prefetches N, the number of buffers Z, and the completion time Δ for the algorithms M_1, and M_2 specified on row 2. For both cases Z_1, and Z_2 specified on row 4, the N and Δ achieved by the KAP algorithm are then given in row 5. The sixth row (Gain 1) shows the relative improvements of KAP — on each problem instance I given in Table 2 — with respect to M_1/M_2 relatively to the lower bound, which is measured by the following formula:

$$\Phi_{KAP}(I) = \Big(\big(MMOpt(I) - KAP(I)\big) / \big(MMOpt(I) - Bound(I)\big) \Big) * 100$$

Similarly, both N and Δ achieved by the SPbP algorithm are then given in row 7. The eighth row (Gain 2) shows the relative improvements of SPbP with respect to M_1/M_2 relatively to the lower bound using the same previous formula $\Phi_{SPbP}(I)$.

The three last rows give for each of the algorithms M_1, M_2, KAP, and SPbP, the ratio of the achieved completion time Δ to the lower bound lb_Δ (given in Table 3). The column Average provides the average gains (%) for all the kernels in the case of Z_1 (Av. 1), respectively, of Z_2 (Av. 2).

As illustrated in Table 3, the number of prefetches, by running the KAP algorithm, is reduced with an average reduction of 57.5% (Z_1), respectively, of 36.8% (Z_2). In contrast, the completion time is increased, with an average increase of 14.9% (Z_1), respectively, of 22% (Z_2). This is due to the absence of overlap between the prefetches and computations in the schedules produced by the KAP algorithm.

Table 3 Numerical results of M_1, M_2, KAP, and SPbP

Kernel N°		1 M_1	1 M_2	2 M_1	2 M_2	3 M_1	3 M_2	4 M_1	4 M_2	5 M_1	5 M_2	Av. 1	Av. 2
MMOpt	Z	20	13	29	21	28	20	18	13	139	96		
	N	395	395	640	640	478	478	1710	1710	3640	3640		
	Δ	907	976	1341	1457	1021	1081	5075	5129	7899	8070		
		Z_1	Z_2	Z_1	Z_2	Z_1	Z_2	Z_1	Z_2	Z_1	Z_2	Z_1	Z_2
KAP	N	298	322	457	517	353	405	1283	1458	2560	2997		
	Δ	1071	1119	1389	1509	1043	1147	6125	6475	6405	7279		
Gain 1	N	56.7	42.6	65.3	43.9	53.4	31.1	33.3	19.6	78.9	47.0	57.5 %	36.8 %
	Δ	−38.1	−28.6	−7.7	−7.0	−4.1	−11.2	−69.3	−85.8	44.5	22.4	−14.9 %	−22.0 %
SPbP	N	298	322	457	517	353	405	1283	1458	2560	2997		
	Δ	795	871	1113	1226	851	947	4629	4916	5849	6789		
Gain 2	N	56.7	42.6	65.3	43.9	53.4	31.1	33.3	19.6	78.9	47.0	57.5 %	36.8 %
	Δ	26.0	21.0	36.9	31.5	32.1	22.7	29.4	13.5	61.1	36.3	37.1 %	25.0 %
Δ_{MMOpt}/lb_Δ		1.90	2.04	1.85	2.01	2.07	2.19	1.42	1.44	1.73	1.77	1.79	1.89
Δ_{KAP}/lb_Δ		2.24	2.34	1.91	2.08	2.11	2.33	1.72	1.81	1.40	1.60	1.87	2.03
Δ_{SPbP}/lb_Δ		1.66	1.82	1.53	1.69	1.72	1.92	1.29	1.38	1.28	1.49	1.49	1.66

In addition, due to the reuse of the KAP algorithm as a subroutine of the SPbP algorithm, the traffic to the external memory is reduced with the same average reduction of 57.5% (Z_1), respectively, of 36.8% (Z_2). In contrast to KAP, minimizing Δ by SPbP leads to a 37.1% (Z_1), respectively, to a 25% (Z_2) decrease in average of the completion time.

In the same way, a comparison between the completion time Δ achieved by each of MMOpt's original algorithms (M_1 and M_2), KAP, and SPbP and the lower bound lb_Δ, is considered. A comparison against the lower bound provides a measure of deviation from optimality. It is used as a performance indicator, and calculated by taking the ratio of Δ to lb_Δ.

As shown in Table 3, for both cases Z_1, and Z_2, the completion time Δ of SPbP algorithm is in average closer to the value of lb_Δ than the different values given by each of the algorithms M_1, M_2, and KAP. It is at most twice the value of lb_Δ. This is achieved by delaying computations as a measure to spare the buffers. This means also that the SPbP algorithm gives a better completion time Δ than the other methods.

We next analyze the performance of the ECM, CGM, E-ECM and E-CGM algorithms. We recall that the number of prefetches equals its lower bound lb_N in the context of the different algorithms under consideration: $N_{ECM} = N_{E-ECM} = N_{CGM} = N_{E-CGM} = lb_N$. However, the number of buffers Z is larger than its lower bound lb_Z ($Z_{ECM} = Z_{CGM} \gg lb_Z$) and the completion time Δ is minimized in the context of both ECM and CGM algorithms. Whereas, Z is minimized and Δ has the the same values as those provided by both ECM and CGM algorithms, which are used as a subroutine in the context of both E-ECM and E-CGM algorithms: $\Delta_{ECM} = \Delta_{E-ECM}$ and $\Delta_{CGM} = \Delta_{E-CGM}$.

Figures 9 and 10 show, respectively, the completion time Δ found by each of the ECM and CGM algorithms and the number of buffers Z provided by each of the E-ECM and E-CGM algorithms for all the non-linear kernels given in Table 2.

As shown in Fig. 9, the value of Δ provided by the CGM algorithm, for the first four kernels, seems much closer to the one found by the ECM algorithm. In contrast, the value Δ found by the ECM algorithm equals its lb_Δ in the case of the fifth kernel.

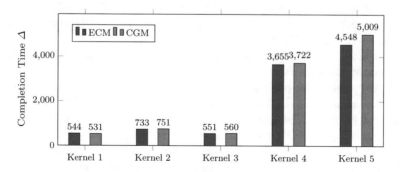

Fig. 9 ECM and CGM results

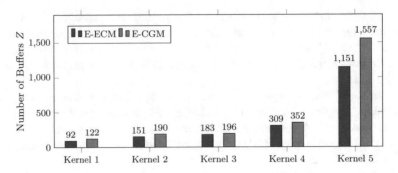

Fig. 10 E-ECM and E-CGM results

Table 4 Comparison of ECM, CGM, E-ECM and E-CGM against bounds

Kernel N°	1	2	3	4	5	Average
Δ_{ECM} / lb_Δ	1.14	1.01	1.11	1.02	1.00	**1.05**
Δ_{CGM} / lb_Δ	1.11	1.08	1.13	1.04	1.10	**1.09**
Z_{E-ECM} / lb_Z	7.07	7.19	9.15	23.76	11.98	**11.83**
Z_{E-CGM} / lb_Z	9.38	9.04	9.80	27.07	16.21	**14.30**

Similarly, Fig. 10 shows that the number of buffers Z found by the E-ECM algorithm seems much better than the one provided by the E-CGM algorithm.

In the same way, Table 4 gives a comparison between Δ, found by each of the ECM and CGM algorithms, and its lower bound lb_Δ, together with a comparison between Z, provided by each of the E-ECM and E-CGM algorithms and its lower bound lb_Z. This comparison is calculated by taking the ratio, respectively, of Δ to lb_Δ and of Z to lb_Z.

As illustrated in Table 4, the completion time Δ provided by the ECM algorithm is very close, more than the one found by the CGM algorithm, to the value of its lower bound lb_Δ. This implies that the lb_Δ is a good lower bound on the completion time Δ for the main 3-PSDPP.

In addition, the number of buffers Z found by both E-ECM and E-CGM algorithms is much larger than its lower bound lb_Z.

We also analyze the performance of both CCM and E-CCM algorithms, which are described in Sects. 6.3.3 and 6.4. Note that the CCM algorithm is used as a subroutine of the E-CCM algorithm. Both CCM and E-CCM algorithms have the same values for N, which equals its lower bounds ($N_{CCM} = N_{E-CCM} = lb_N$), and for Δ. However, the number of buffers provided by the CCM algorithm is larger than its lower bound lb_Z ($Z_1 \gg lb_Z$), which equals its number of required input tiles $\mathcal{X}'$, and it is minimized in the context of the E-CCM algorithm.

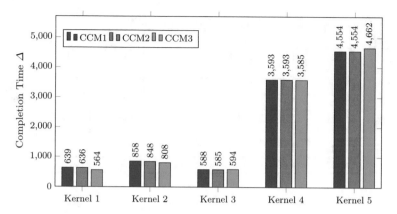

Fig. 11 CCM and E-CCM: results of Δ

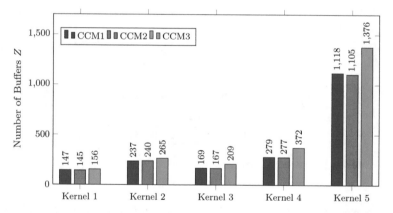

Fig. 12 E-CCM: results of Z

Figures 11 and 12 show the completion time Δ found by both CCM and E-CCM algorithms together with the number of buffers Z provided by the E-CCM algorithm — for the three ideas (CCM1, CCM2 and CCM3) proposed to sequence the partition of classes found in the first step of this algorithm — for all the non-linear kernels given in Table 2.

As shown in these figures, all the values of Δ are close to each other. This means that the minimization of the completion time can be done with a large variety of methods that give almost the same value of Δ. In contrast, the value of Z provided by the E-CCM algorithm for both CCM1 and CCM2 ideas, seems much better than the one found in the context of the CCM3 proposition.

In the same way, Table 5 gives a comparison between both Δ and Z values, found by both CCM and E-CCM algorithms for each one of the three ideas (CCM1, CCM2 and CCM3), and its lower bounds lb_Δ and lb_Z.

Table 5 Comparison between CCM and E-CCM against bounds

Kernel Nº		1	2	3	4	5	Average
Δ_{CCM} / lb_Δ	CCM1	1.33	1.18	1.19	1.01	1.00	**1.14**
	CCM2	1.33	1.17	1.18	1.01	1.00	**1.13**
	CCM3	1.14	1.11	1.20	1.01	1.02	**1.09**
Z_{E-CCM} / lb_Z	CCM1	11.30	11.28	8.45	21.46	11.64	**12.82**
	CCM2	11.15	11.42	8.35	21.30	11.51	**12.74**
	CCM3	12.00	12.61	10.45	28.61	14.33	**15.60**

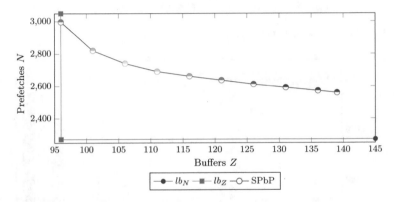

Fig. 13 E-SPbP - Energy versus Area

As shown in this table, all the values of Δ are very close to the value of its lower bound lb_Δ. However, the value of Z provided by each one of the three ideas is much larger than its lower bound lb_Z.

We finally analyze the performance of the E-SPbP algorithm, which is described in Sect. 6.4. Figures 13, 14 and 15 consider kernel 5 which is the biggest one tested with more than 7000 input tiles. By varying the value of Z from $Z_2 = 96$ to $Z_1 = 139$ (note that the exploration of the whole interval is not executed), Fig. 11 presents for each pair of criteria — (N, Z), (Δ, Z) and (Δ, N) — a set of ten different solutions, giving three different views on the same solution set in order to let MMOpt's user to make a more informed decision-making about his favorite compromise solution.

As shown in both Figs. 13 and 14, when we increase the number of buffers Z, the values of both N and Δ provided by the E-SPbP algorithm decrease. This means that both energy and time decrease in the same way with respect to the increase of the parameter area. However, Fig. 15 shows the simultaneous increase of both N and Δ parameters. As shown in this figure, when the number of buffers is small, the E-SPbP algorithm gives a large number of both N and Δ. In contrast, to find a good value for both N and Δ, the number of buffers Z has the big value. This means that by varying

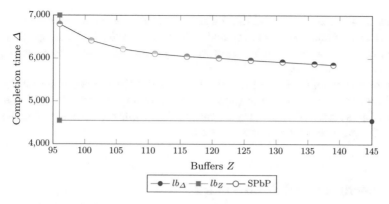

Fig. 14 E-SPbP's solutions for Kernel 5 - Time versus Area

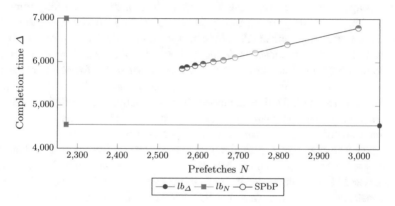

Fig. 15 E-PSbP's solutions for Kernel 5 - Time versus Energy

the embedded memory area (Z), new trade-offs between the energy consumption (N) and the computing time (Δ) can be then reached.

In summary, numerical experiments, which are conducted on a set of benchmarks from real-life non-linear image processing kernels as used by Mancini et al. [14, 15], show that the schedules produced by each of our proposed algorithms give good results for the different optimization criteria (N, Z, Δ).

According to contradictory needs of MMOpt's user — which are the low cost, the low energy and the high performance of the produced circuit —, his choice depends on his decision about which criterion he first wants minimized. In fact, if MMOpt's user wants to optimize a one criterion at a time, he can then choose his preferred algorithm from the set of polynomial heuristics described in Sect. 6, which were proposed for solving each one of the mono-objective sub-problems described in Sect. 3.3. However, if he wants to optimize two or three criteria simultaneously, new trade-offs between the embedded memory area, the computing time and the energy consumption can be then reached:

- The E-ECM, E-CGM and E-CCM algorithms, the provided number of prefetches N has an optimal value, which equal its lower bound lb_N, and a completion time Δ close to its lower bound lb_Δ. However, they use a large value of Z, which equals in average $13.45 * lb_Z$.
- In the context of the E-SPbP algorithm, by varying the number of buffers Z, a set of compromise solutions was proposed that allows us to provide useful ones for MMOpt's user. In this case, he can choose his favorite compromise solution from the design space that he wants or needs.

8 Conclusion and Future Work

In this chapter, we have investigated a new multi-objective scheduling optimization problem, 3-PSDPP, arising in the context of embedded vision systems. This optimization challenge reflects the efficient operation of the TPUs produced by the MMOpt tool, which is proposed by Mancini et al. [14, 15] as an architectural solution that creates an ad-hoc generator of memory hierarchies suited for non-linear access patterns to alleviate the problematic of memory management ("Memory Wall"). We demonstrated how optimization techniques from OR help us build efficient solutions to enhance the electronic characteristics of the circuits produced by MMOpt, such as production cost, energy consumption and performance.

We gave a specific mathematical model to formulate this problem, as well as several sub-problems (mono- and bi-objective) of interest, proved some lower bounds, and analyzed the complexity of its mono-objective sub-problems. Several polynomial constructive heuristics were proposed to solve the main 3-PSDPP together with its sub-problems. To evaluate their effectiveness, numerical experiments were conducted on the same real-world data set as used by Mancini et al. [14, 15]. A comparison against MMOpt's original algorithms shows a very significant improvement in terms of a reduction of both the amount of transferred data and the total completion time.

An interesting area for further research may be the improvement of the proposed methods, the complexity analysis of the MCT-PSDPP sub-problem, in which the completion time Δ is to be minimized, and the development of an exact optimization procedure (e.g., *Integer Linear Programming*: ILP) for solving the NP-Complete 3-PSDPP sub-problems. It would also seem interesting to develop specific multi-objective optimization techniques to solve our main 3-PSDPP — such as the *Multi-Objective Evolutionary Algorithms* (MOEAs). Additional research is also required to use our own heuristics approaches to solve other variants of the ToSP problem.

References

1. Amaya, J., Cotta, C., Fernández, A.: A memetic algorithm for the tool switching problem. In: Hybrid metaheuristics, pp. 190–202. Springer (2008)
2. Applegate, D., Bixby, R., Cook, W., Chvátal, V.: On the solution of traveling salesman problems. Rheinische Friedrich-Wilhelms-Universität Bonn (1998)
3. Balakrishnan, N., Chakravarty, A.: Opportunistic retooling of a flexible machine subject to failure. Naval Res. Logist. **48**(1), 79–97 (2001)
4. Bard, J.: A heuristic for minimizing the number of tool switches on a flexible machine. IIE Trans. **20**(4), 382–391 (1988). doi:10.1080/07408178808966195
5. Bellas, N., Chai, S., Dwyer, M., Linzmeier, D.: Real-time fisheye lens distortion correction using automatically generated streaming accelerators. In: 17th IEEE Symposium on Field Programmable Custom Computing Machines, FCCM'09, pp. 149–156 (2009). doi:10.1109/FCCM.2009.16
6. Bertossi, A.: The edge hamiltonian path problem is NP-complete. Inf. Process. Lett. **13**(4), 157–159 (1981)
7. Błażewicz, J., Finke, G.: Scheduling with resource management in manufacturing systems. Eur. J. Oper. Res. **76**(1), 1–14 (1994)
8. Catanzaro, D., Gouveia, L., Labbé, M.: Improved integer linear programming formulations for the job sequencing and tool switching problem. Eur. J. Oper. Res. **244**(3), 766–777 (2015)
9. Crama, Y.: Combinatorial optimization models for production scheduling in automated manufacturing systems. Eur. J. Oper. Res. **99**(1), 136–153 (1997)
10. Crama, Y., Kolen, A., Oerlemans, A., Spieksma, F.: Minimizing the number of tool switches on a flexible machine. Int. J. Flex. Manuf. Syst. **6**(1), 33–54 (1994)
11. Hadj Salem, K., Kieffer, Y., Mancini, M.: Formulation and practical solution for the optimization of memory accesses in embedded vision systems. In: Proceedings of the 2016 Federated Conference on Computer Science and Information Systems, FedCSIS 2016, Gdańsk, Poland, 11–14 Sept 2016, pp. 609–617 (2016). doi:10.15439/2016F124
12. Konak, A., Kulturel-Konak, S.: An ant colony optimization approach to the minimum tool switching instant problem in flexible manufacturing system. In: 2007 IEEE Symposium on Computational Intelligence in Scheduling (2007)
13. Laporte, G., Salazar-Gonzalez, J., Semet, F.: Exact algorithms for the job sequencing and tool switching problem. IIE Trans. **36**(1), 37–45 (2004). doi:10.1080/07408170490257871
14. Mancini, S., Rousseau, F.: Enhancing non-linear kernels by an optimized memory hierarchy in a high level synthesis flow. In: Proceedings of the Conference on Design, Automation and Test in Europe, pp. 1130–1133. EDA Consortium (2012)
15. Mancini, S., Rousseau, F.: Optimisation d'accélérateurs matériels de traitement par incorporation d'un gestionnaire de données et de contrôle dans un flot de HLS. In: Conférence en Parallélisme, Architecture et Système (2013)
16. Privault, C., Finke, G.: Modelling a tool switching problem on a single nc-machine. J. Intell. Manuf. **6**(2), 87–94 (1995). doi:10.1007/BF00123680
17. Tang, C., Denardo, E.: Models arising from a flexible manufacturing machine, part i: minimization of the number of tool switches. Oper. Res. **36**(5), 767–777 (1988)
18. Thornton, A., Sangwine, S.: Log-polar sampling incorporating a novel spatially variant filter to improve object recognition. Sixth Int. Conf. Image Process. Appl. **2**, 776–779 (1997). doi:10.1049/cp:19971001
19. Viola, P., Jones, M.: Robust real-time face detection. Int. J. Comput. Vis. **57**(2), 137–154 (2004)

Analysis and Experimental Study of Heuristics for Job Scheduling Reoptimization Problems

Elad Iwanir and Tami Tamir

Abstract Many real-life applications involve systems that change dynamically over time. Thus, throughout the continuous operation of such a system, it is required to compute solutions for new problem instances, derived from previous instances. Since the transition from one solution to another incurs some cost, a natural goal is to have the solution for the new instance close to the original one (under a certain distance measure). We study reoptimization problems arising in scheduling systems. Formally, due to changes in the environment (out-of-order or new machines, modified jobs' processing requirements, etc.), the schedule needs to be modified. That is, jobs might be migrated from their current machine to a different one. Migrations are associated with a cost – due to relocation overhead and machine set-up times. In some systems, a migration is also associated with job extension. The goal is to find a good modified schedule, with a low transition cost from the initial one. We consider reoptimization with respect to the classical objectives of minimum makespan and minimum total flow-time. We first prove that the reoptimization variants of both problems are NP-hard, already for very restricted classes. We then develop and present several heuristics for each objective, implement these heuristics, compare their performance on various classes of instances and analyze the results.

1 Introduction

Reoptimization problems arise naturally in dynamic scheduling environments, such as manufacturing systems and virtual machine managers. Due to changes in the environment (out-of-order or new resources, modified jobs' processing requirements, etc.), the schedule needs to be modified. That is, jobs may be migrated from their current machine to a different one. Migrations are associated with a cost due to relocation overhead and machine set-up times. In some systems, a migration is also

E. Iwanir · T. Tamir (✉)
School of Computer Science, The Interdisciplinary Center, Herzliya, Israel
e-mail: tami@idc.ac.il

E. Iwanir
e-mail: eladiw@gmail.com

© Springer International Publishing AG 2018

S. Fidanova (ed.), *Recent Advances in Computational Optimization*,
Studies in Computational Intelligence 717, DOI 10.1007/978-3-319-59861-1_12

associated with job extension. The goal is to find a good modified schedule, with a low transition cost from the initial one.

This work studies the reoptimization variant of two classical scheduling problems in a system with identical parallel machines: (*i*) minimizing the total flow-time (denoted in standard scheduling notation by $P||\Sigma C_j$ [15]), and (*ii*) minimum makespan (denoted by $P||C_{max}$).

The minimum total flow-time problem for identical machines can be solved efficiently by the simple greedy Shortest Processing Time algorithm (SPT) that assigns the jobs in non decreasing order by their length. The minimum makespan problem is NP-hard, and has several efficient approximation algorithms, as well as a polynomial-time approximation scheme [17]. These algorithms, as many other algorithms for combinatorial optimization problems, solve the problem from scratch, for a single arbitrary instance without having any constraints or preferences regarding the required solution - as long as they achieve the optimal objective value. However, many of the real-life scenarios motivating these problems involve systems that change dynamically over time. Thus, throughout the continuous operation of such a system, it is required to compute solutions for new problem instances, derived from previous instances. Moreover, since the transition from one solution to another consumes energy (used for the physical migration of the job, for warm-up or set-up of the machines, or for activation of the new machines), a natural goal is to have the solution for the new instance close to the original one (under certain distance measure). Solving a reoptimization problem involves two challenges:

1. Computing an optimal (or close to the optimal) solution for the new instance.
2. Efficiently converting the current solution to the new one.

Each of these challenges, even when considered alone, gives rise to many theoretical and practical questions. Obviously, combining the two challenges is an important goal, which shows up in many applications.

1.1 Problem Description

An instance of our problem consists of a set J of n jobs and a set M_0 of m_0 identical machines. Denote by p_j the processing time of job j. An initial schedule S_0 of the jobs is given. That is, for every machine, it is specified what are the jobs it processes. At any time, a machine can process at most one job and a job can be processed by at most one machine. We consider the scenario in which a change in the system occurs. Possible changes include addition or removal of machines and/or jobs, as well as modification of processing times of jobs in J. We denote by M the set of machines in the modified instance and let $m = |M|$.

Our goal is to suggest a new schedule, S, for the modified instance, with good objective value and small transition cost form S_0. Assignment of a job to a different machine in S_0 and S is denoted *migration* and is associated with a cost. Formally, we are given a *price list* $\theta_{i,i',j}$, such that it costs $\theta_{i,i',j}$ to migrate job j from machine i to machine i'. Moreover, in some systems job migrations are also associated with an extension of the job's processing time. Formally, in addition to the transition costs, we are given a *job-extension penalty list* $\delta_{i,i',j} \geq 0$, such that the processing time of job j is extended to $p_j + \delta_{i,i',j}$ when it is migrated from machine i to machine i'.

For a given schedule, let C_j denote the *flow-time* (also known as 'completion time') of job j, that is, the time when the process of j completes. The *makespan* of a schedule is defined as the maximal completion time of a job, that is, $C_{max} = \max_j C_j$.

While the schedule does not specify the internal order of jobs assigned to a machine, we assume throughout this work that jobs assigned to a specific machine are always processed in SPT-order (Shortest Processing Time), that is, $p_1 \leq p_2 \leq \ldots$. For a given set of jobs, SPT-order is known to achieve the minimal possible total flow-time, min $\sum_j C_j$ [9, 25]. Clearly, the internal order has no effect on the makespan.

Given S_0, J and M, the goal is to find a good schedule for J that is close to the initial schedule S_0. We consider two problems:

1. Rescheduling to an optimal schedule using the minimal possible transition cost.
2. Given a budget B, find the best possible modified schedule that can be achieved without exceeding the budget B.

If the modification includes machines' removal, and the budget is limited, then we assume that a feasible solution exists. That is, the budget is at least the cost of migrating the jobs on the removed machines to some machine. Formally, $B \geq \sum_{j \text{ on } i \in M_0 \setminus M} \min_{i' \in M} \theta_{i,i',j}$.

Applications: Our reoptimization problems arise naturally in manufacturing systems, where jobs may be migrated among production lines. Due to unexpected changes in the environment (out-of-order or new machines, timetables of task processing, etc.), the production schedule needs to be modified. Rescheduling tasks involves energy-loss due to relocation overhead and machine set-up times. In fact, our work is relevant to any dynamic scheduling environment, in which migrations of jobs are allowed though associated with an overhead caused due to the need to handle the modification and to absorb the migrating jobs in their new assignment.

With the proliferation of cloud computing, more and more applications are deployed in the data centers. Live migration is a common process in which a running virtual machine (VM) or application moves between different physical machines without disconnecting the client or application [8]. Memory, storage, and network connectivity of the virtual machine are transferred from the original host machine to the destination. Such migrations involve a warm-up phase, and a memory-copy phase. In pre-copy memory migration, the Hypervisor typically copies all the memory pages from source to destination while the VM is still running on the source. Alternatively, in post-copy memory migration the VM is suspended, a minimal subset of the execution state of the VM (CPU state, registers and, optionally, non-pageable memory) is transferred to the target, and the VM is then resumed at the target. Live migration is performed in several VM managers such as Parallels Virtuozzo [21] and Xen [27]. Sequential processing of jobs that might be migrated among several processors is performed also in several implementations of MapReduce (e.g., [4]), and in RPC (Remote Procedure Call) services, in which virtual servers can be temporarily rented [5].

1.2 Related Work

The work on reoptimization problems started with the analysis of dynamic graph problems (see e.g. [12, 26]). These works focus on developing data structures supporting update and query operations on graphs. A different line of research, deals with the computation of a good solution for an NP-hard problem, given an optimal solution for a close instance. Among the problems studied in this setting are TSP [1, 6] and Steiner Tree on weighted graphs [13].

The paper [24] suggests the framework we adopt for this work, in which the solution for the modified instance is evaluated also with respect to its difference from the initial solution. Formally, an algorithm $\mathscr{A}$ is an (r, ρ)-reapproximation algorithm if it achieves a ρ-approximation for the optimization problem, while paying a transition cost that is at most r times the minimum required for solving the problem optimally. For this definition, the paper [24] characterizes classes of combinatorial reoptimization problems that obey a fully polynomial time $(1 + \varepsilon_1, 1 + \varepsilon_2)$-reapproximation schemes, and suggest reapproximation algorithms for the metric k-Center problem, and for subset-selection problems. This framework of reapproximation is in use also in [23], to analyze algorithms for data placement in storage area network.

Job scheduling reoptimization problems with respect to the total flow-time objective was studied in [2]. The paper includes optimal algorithms for the problem of achieving an optimal solution using the minimal possible transition cost. For the modification of machines' addition, where jobs are only allowed to migrate to new machines, an optimal algorithm for achieving the best possible schedule using a given limited budget is also presented. The paper [3] analyze the problem as a load balancing game: each job corresponds to a selfish agent, whose cost depends on the load on the machine it is assigned. Thus, when machines are added, jobs have an incentive to migrate to the new unloaded machines. When machines are removed, the jobs assigned to them must be reassigned. Migration is associated with a cost, taken into account by the agents. The paper [3] analyzes the stability of these games and the inefficiency caused due to the agents' selfish behaviour.

Lot of attention, in both the industry and the academia, is given recently to the problem of minimizing the overhead associated with migrations (see e.g., [8, 16]). Using our notations, this refers to minimizing the transition costs and the job-extension penalties associated with rescheduling a job. Our work focuses in determining the best possible schedule given these costs.

1.3 Our Contribution

Our study includes both theoretical and comprehensive experimental results. We consider reoptimization with respect to the two classical objectives of minimum makespan and minimum total flow-time, and distinguish between instances with unlimited and limited budget. Our results are presented in Table 1. For completeness, we include in the table previous results regarding the minimum total flow-time with unlimited budget [2].

Table 1 Summary of our results

Objective Function	Optimal Reschedule Using Minimum Budget			Best Reschedule Using Given Limited Budget		
	NP-hard	Optimal Solution	Heuristics	NP-hard	Optimal Solution	Heuristics
$\min C_{max}$	Yes	Branch and Bound	• LPT-based greedy • Loads-based greedy • Genetic algorithm	Yes	Branch and Bound	• LPT-based greedy • Loads-based greedy • Genetic algorithm
$\min \sum_j C_j$	No [2]	[2]	–	Yes	Branch and Bound	• Greedy reversion • Cyclic reversion • SPT-like • Genetic algorithm

We first analyze the computational complexity status of these problems. While the hardness result for the minimum makespan problem is straightforward, the hardness for the minimum total flow-time problem with limited budget is complex and a bit surprising - given that the minimum flow-time problem is solvable even on unrelated machines [7, 18], and that the dual variant of achieving an optimal reschedule using minimum budget is also solvable [2]. Our hardness results, given in Sect. 2 are valid already for very restricted classes, with a single added machine and no job-extension penalties. In Sect. 3 we consider the minimum makespan reoptimization problem assuming that the modification consists of machines' removal and the budget is the minimal possible, that is, sufficient only for reassignment of the jobs on the removed machines. We present tight bounds on our ability to compete against an optimal algorithm with unlimited budget, and present a polynomial time approximation scheme where the approximation ratio is measured with respect to a solution achieved using the same minimal budget.

The second part of the paper includes an experimental study of the problem. In order to be able to evaluate our heuristics against an optimal solution, we develop and implement an optimal solver based on *branch and bound* technique. Naturally, the solver could not handle very large instances, but we were able to run it against small but diverse instances to compare the different heuristics to the optimum. For example, problems instances with 4 machines and 20 jobs were easily solved.

We then present several heuristics for each objective function. Some of the heuristics distinguish between modifications that involve addition or removal of machines. For both objectives we also developed and applied a genetic algorithm [10, 22]. All the heuristics were implemented, their performance on various classes of instances have been compared, and the results were analyzed. Our experimental study concludes the paper.

2 Computational Complexity

In this section we analyze the computational complexity of reoptimization scheduling problems. We distinguish between the minimum makespan and the minimum total flow problems, as well as between the problem of finding an optimal solution using minimum budget and finding the best solution that can be obtained using limited budget.

We use the following notations: For a multiset $A = \{a_1, a_2, \ldots, a_{|A|}\}$ of integers, let $MAX(A) = \max_{j=1}^{|A|} a_j$ and $SUM(A) = \sum_{j=1}^{|A|} a_j$. Also, let $\mathbf{A}$ denote the vector consisting of the elements of A in non-decreasing order and define $SPT(A) = \mathbf{A} \cdot (|A|, \ldots, 2, 1)$. For example, for $A = \{5, 3, 1, 5, 8\}$ it holds that $SUM(A) = 22$, $\mathbf{A} = (1, 3, 5, 5, 8)$ and $SPT(A) = (1, 3, 5, 5, 8) \cdot (5, 4, 3, 2, 1) = 5 + 12 + 15 + 10 + 8 = 50$. Note that $SPT(A)$ is the value of an optimal solution for the minimum total-flow problem on a single machine for an instance consisting of $|A|$ jobs with lengths in A.

For the minimum makespan reoptimization problem, our result is not surprising - the classical load-balancing problem $P||C_{max}$ is known to be NP-complete even with no transition costs or extensions. For completeness we show that this hardness carry over to the simplest class of the reoptimization variant.

Theorem 1 *The minimum makespan reoptimization problem is NP-complete even with a single added machine, unlimited budget, and no job-extension penalty.*

Proof The reduction is from the Partition problem: given a set $A = \{a_1, a_2, \ldots, a_n\}$ of integers, whose total sum is $2B$, the goal is to decide whether A has a subset $A' \subset A$ such that $SUM(A') = SUM(A \setminus A') = B$. The Partition problem is known to be NP-complete [14]. Given an instance of Partition $A = \{a_1, a_2, \ldots, a_n\}$, let $Z = SPT(A)$. We construct the following instance for the reoptimization problem: The initial schedule S_0 consists of a single machine and n jobs having lengths corresponding to the Partition elements, that is, for $1 \leq j \leq n$ let $p_j = a_j$. The modification is simply an addition of one machine. Clearly, a balanced solution with makespan B exists if and only if A has a partition.

The analysis of the minimum total flow problem is more involved. The corresponding classical optimization problem $P|| \sum C_j$ is known to be solvable in polynomial time by the simple SPT algorithm. For the reoptimization problem, an efficient optimal algorithm for finding an optimal solution using minimum budget is presented in [2]. The algorithm is based on a reduction to a minimum weighted perfect matching in a bipartite graph. This reduction cannot be generalized to consider instances with limited budget, and the complexity status of the problem of finding the best solution that can be obtained using limited budget remains open in [2]. We show that this problem is NP-complete, even with no job-extension penalties and a single added machine.

Our proof refers to the compact representation of the problem. In a compact representation, the jobs assigned on each machine are given by a set of pairs $\langle p_j, n_j \rangle$, where n_j is the number of jobs of length p_j assigned on the machine. We first prove two simple observations:

Observation 2 *Let A' be a subset of A then $SPT(A') + SPT(A \setminus A') < SPT(A)$.*

Proof By definition, $SPT(A) = \mathbf{A} \cdot (|A|, \ldots, 2, 1)$. That is, every element is multiplied by its rank in the sorted list. Clearly, for every element in A', its rank in A' is not higher than its rank in A. Similarly, for every element in $A \setminus A'$, its rank in $A \setminus A'$ is not higher than its rank in A.

Given a multiset A, and an integer Z, let $x > MAX(A)$. Extend A to a multiset A^* by adding to it Z elements of value x.

Observation 3 $SPT(A^*) = \frac{Z(Z+1)}{2}x + Z \cdot SUM(A) + SPT(A)$.

Proof Assume w.l.o.g., that $A = \{a_1, a_2, \ldots, a_n\}$, where $a_1 \leq a_2 \leq \ldots \leq a_n$. That is, $\mathbf{A^*} = (a_1, a_2, \ldots, a_n, x, x, \ldots, x)$ with x repeated Z times in the suffix of $\mathbf{A}$.

By definition, $SPT(A^*) = (a_1, a_2, \ldots, a_n, x, x, \ldots, x) \cdot (Z + n, Z + n - 1, \ldots, 1)$. This dot-product can be divided into two parts to get $SPT(A^*) = (a_1, a_2, \ldots, a_n) \cdot (Z + n, Z + n - 1, \ldots, Z + 1) + (x, x, \ldots, x) \cdot (Z, Z - 1, \ldots, 1)$. It can now be easily verified that the left component equals $Z \cdot SUM(A) + SPT(A)$ and the right component equals $\frac{Z(Z+1)}{2} x$.

Using the above observations, we are now ready to prove the hardness result.

Theorem 4 *The minimum total flow-time reoptimization problem with limited budget is NP-complete even with a single added machine, and no job-extension penalty.*

Proof The reduction is from the Partition problem: given a set $A = \{a_1, a_2, \ldots, a_n\}$ of integers, whose total sum is $2B$, the goal is to decide whether A has a subset $A' \subset A$ such that $SUM(A') = SUM(A \setminus A') = B$. The Partition problem is known to be NP-complete [14].

Given an instance of Partition $A = \{a_1, a_2, \ldots, a_n\}$, whose total sum is $2B$, let $Z = SPT(A)$. We construct the following instance for the reoptimization problem: The initial schedule, S_0, consists of a single machine and $n + Z$ jobs. The first n jobs correspond to the Partition elements, that is, for $1 \leq j \leq n$ let $p_j = a_j$. Each of the additional Z *dummy jobs* have length x for some $x > B$. Assume that one machine is added, and that the transition cost of migrating job j from the initial machine to the new machine is p_j. Assume also that the budget is B.

Since the budget is B and each dummy jobs have length more than B, none of these jobs can be migrated to the new machine. Thus, a modified schedule S is characterized by a subset $A' \subset A$ of jobs corresponding to the partition elements, whose total length is at most B. These jobs are migrated to the new machine and assigned in SPT order on it. The remaining jobs are assigned in SPT order on the initial machine. Since x is larger than any a_i, the Z jobs of length x will be assigned after the jobs corresponding to $A \setminus A'$.

Finally, we note that the reduction is polynomial. Calculating $SPT(A)$ requires sorting and is performed in time $O(n\log n)$. The instance constructed in the reduction consists of $n + Z$ jobs where $Z = SPT(A)$. The compact representation of this instance includes at most $n + 1$ different pairs, where all $\langle p_j, n_j \rangle$ values are polynomial in n.

claim The minimum total flow-time in an optimal modified schedule is less than $\frac{Z(Z+1)}{2} x + (B + 1)Z$ if and only if a partition of A exists.

Proof Assume that a partition exists and let A' be a subset of A such that $SUM(A') = SUM(A \setminus A') = B$. Consider the modified schedule in which jobs corresponding to the elements of A' migrate to the new machines. Since the transition cost of job j equals p_j, and $SUM(A') = B$, the budget B exactly fits this modification.

Denote by A^* the multiset produced by adding Z elements of value x to $A \setminus A'$. Note that A^* is exactly the multiset corresponding to job lengths that remain on the initial machine. By Observation 3, the total flow-time of the jobs that remain on the first machines is $SPT(A^*) = \frac{Z(Z+1)}{2} x + Z \cdot SUM(A \setminus A') + SPT(A \setminus A')$. Also, the total flow-time of the jobs that migrate to the new machine is $SPT(A')$. Summing the jobs' flow-time on the two machines, we get that the total flow-time is $\frac{Z(Z+1)}{2}x + Z \cdot SUM(A \setminus A') + SPT(A \setminus A') + SPT(A')$. By Observation 2, $SPT(A \setminus A') + SPT(A') < SPT(A) = Z$. Also, since A' defines a Partition, it holds that $SUM(A \setminus A') = B$. We get that the total flow-time is less than $\frac{Z(Z+1)}{2}x + ZB + Z = \frac{Z(Z+1)}{2}x + (B+1)Z$.

For the other direction of the reduction proof, assume that a partition of A does not exist. This implies that the budget B cannot be fully exploited and thus, the set A'' of jobs migrating to the new machine includes jobs of total length strictly less than B. Thus, the set $A \setminus A''$ of jobs that remain on the initial machine has total length at least $B + 1$. By Observation 3, the total flow-time of the jobs that remain on the initial machines is at least $\frac{Z(Z+1)}{2}x + Z \cdot (B+1) + SPT(A \setminus A'')$. The total flow-time of the jobs on the new machine is $SPT(A'')$. Since $SPT(A \setminus A'') + SPT(A'') > 0$, we get that if a partition does not exist, then the total flow-time is more than $\frac{Z(Z+1)}{2}x + Z \cdot (B+1)$ as required.

The above claim completes the hardness proof.

Remark It is possible that two multisets A, B will have the same cardinality, and that $SUM(A) > SUM(B)$ while $SPT(A) < SPT(B)$. For example, for $A = \{1, 2, 10\}$ and $B = \{3, 4, 5\}$, we have $|A| = |B| = 3$, $SUM(A) = 13 > 12 = SUM(B)$ while $SPT(A) = 15 < 24 = SPT(B)$. This emphasis an additional challenge of the reoptimization problem: an optimal solution may not use the whole budget. Such anomalies also explains why our hardness proof cannot be a simple reduction from the subset-sum problem, and the dummy jobs are required.

3 Approximation Algorithms for Machines' Removal and Minimum Budget

In this section we consider the minimum makespan reoptimization problem assuming that the modification consists of machines' removal and the budget is the minimal possible, that is, sufficient only for reassignment of the jobs on the removed machines. Recall that M_0 is the set of machines in the initial instance, M is the set of machines in the modified instance, and let $M' = M_0 \setminus M$ be the set of removed machines. Denote $m_0 = |M_0|$, $m = |M|$, and $m' = |M'|$. We assume that the cost and the extension penalty for migrating a job j is independent of the target machine, that is, for all j, and all i', it holds that $\theta_{i,i',j} = \theta_{i,j}$ and $\delta_{i,i',j} = \delta_{i,j}$. The given budget is exactly $B_{min} = \sum_j$ on machine $_{i \in M'} \theta_{i,j}$. Denote by $OPT(B_{min})$ the minimum makespan that can be achieved using this minimal budget, and let $OPT(B_\infty)$ denote the minimum makespan that can be achieved using unlimited budget. We first analyze the ratio $OPT(B_{min})/OPT(B_\infty)$. Note that $OPT(B_\infty)$ refers to a schedule that can assign 'from scratch' all the jobs in J, independent of their original machines. Clearly, if S_0 is far from being optimal, then we cannot expect to be competitive against unlimited budget. The following theorem provides tight analysis of the minimal r, such that if S_0 is an r-approximation, then $OPT(B_{min})/OPT(B_\infty) \leq r$. Let C_{max} and $C^*(S_0)$ denote the makespan and the minimal possible makespan respectively, of the initial instance S_0.

Theorem 5 *For every instance, and every number of initial and removed machines, if $C_{max}(S_0) \leq (2 - \frac{1}{m})C^*(S_0)$, then $OPT(B_{min})/OPT(B_\infty) \leq 2 - \frac{1}{m}$.*

Proof We show that this bound is achieved by a simple List-Scheduling method. Specifically, the jobs on M' are considered in arbitrary order; each job is assigned on a machine in M with minimal load. Ties are broken arbitrarily.

For every job j assigned to machine $i \in M'$, let $p'_j = p_j + \delta_{i,j}$. Note that p'_j is the processing time of job j after its migration - independent of the target machine. Let $C_{LS}(S)$ be the makespan of the resulting assignment, and let J_k, of length p_k (including the extension penalty), be the job determining the makespan. If J_k was on M before the modification, then the makespan was not increased due to the migrating jobs, and $C_{LS}(S) = C_{max}(S_0) \leq (2 - \frac{1}{m})C^*(S_0) \leq (2 - \frac{1}{m})C^*(S)$. The last inequality follows from the fact that $C^*(S_0) \leq C^*(S)$, as S is a schedule of the same set of jobs on fewer machines.

Assume J_k was migrated from some removed machine. As all machines are busy at time $C_{LS}(S) - p_k$, it holds that $\sum_j p_j \geq (m-1)(C_{LS}(S) - p_k) + C_{LS}(S)$. Therefore, $C_{LS} \leq \frac{1}{m}\sum_j p_j + p_k \frac{m-1}{m}$. Clearly, $C^*(S) \geq \frac{1}{m}\sum_j p_j$, and $C^*(S) \geq p_k$. Therefore, $C_{LS}(S) \geq OPT(B_{min}) \geq OPT(B_\infty)$.

The above analysis is tight as demonstrated in Fig. 1. This tight bound is identical to the tight bound of the analysis of List-Scheduling. For any number of machines m_0, the initial schedule is depicted in Fig. 1a. Assume $m' = 1$, that is, $m_0 = m + 1$. Each of the remaining m machine is assigned $m - 1$ unit-length jobs, and the removed machine is assigned one job of length m, that should be migrated. If the budget is sufficient only for migrating the job on the removed machine, then the schedule depicted in Fig. 1b is optimal and $OPT(B_{min}) = 2m - 1$. On the other hand, if additional jobs may be migrated, then, as shown in Fig. 1c, $OPT(B_\infty) = m$. The ratio is $2 - \frac{1}{m}$.

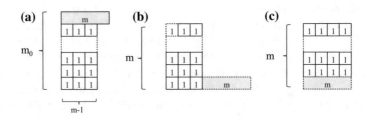

Fig. 1 **a** An initial schedule, **b** An optimal reschedule with minimal budget. $C_{max} = 2m - 1$, **c** An optimal reschedule with unlimited budget, $C_{max} = m$

We note that this example can be generalized to any number m' of removed machines – by assigning them jobs of length $\varepsilon \to 0$ in the initial schedule. This example also demonstrates that even if S_0 is optimal, a minimal budget cannot guarantee approximation ratio better than $2 - \frac{1}{m}$.

The above bound suggests that the performance of a reapproximation algorithm should not be measured compared to an algorithm with unlimited budget, but compared to an algorithm with unlimited *computational power* and the same budget. We present a polynomial time approximation scheme (PTAS) for this measure. Formally, for any instance S_0, and any given $\varepsilon > 0$, our algorithm calculates a reassignment S of the jobs on the removed machines, using budget B_{min}, such that $C_{max}(S) \leq (1 + \varepsilon)OPT(B_{min})$.

We reduce the reapproximation problem to the minimum makespan problem with processing set restrictions [20]. In this scheduling problem, each job j is associated with a subset $P(j)$ of the machines on which it can be processed. The paper [11] presents polynomial time approximation schemes for several variants of this problem, in particular, the one we use below. The idea of our algorithm is to glue, for every machine $i \in M$, all the jobs assigned to machine i in S_0 into a single job. This job must be assigned to its original machine in S_0, while the migrating jobs are not limited in their processing set.

Clearly, every valid schedule of the instance constructed in Step 3 corresponds to a modified schedule in which the only migrating jobs are the jobs from the removed machines. We can therefore use as a black-box the PTAS of [11] to conclude:

Theorem 6 *For any $\varepsilon > 0$ and any instance of removed machines, Algorithm 1 uses budget B_{min} and produces an assignment S such that $C_{max}(S) \leq (1 + \varepsilon) OPT(B_{min})$.*

Algorithm 1 PTAS for machines' removal and minimum budget

1: For every job assigned to a removed machine $i \in M'$, let $p'_j = p_j + \delta_{i,j}$.

2: Let $\ell_0(i)$ be the load on machine $i \in M$ in S_0, that is, the total length of jobs assigned to machine i before the modification.

3: Construct an instance for the minimum makespan problem with m identical machines and the following jobs and processing restrictions:

   - Every job j assigned to a removed machine i in S_0 contributes one job of length p'_j, for which $P(j) = M$.
   - Every machine $i \in M$ contributes one job of length $\ell_0(i)$, for which $P(j) = \{i\}$.

4: Run a PTAS for the minimum makespan problem with processing set restrictions [11] on the resulting instance, and the given parameter $\varepsilon > 0$.

---

4 Optimal Algorithms

4.1 A Brute-Force Solver Based on Branch and Bound

Our brute-force solver was designed to utilize high performance multi-core machines in order to find optimal solutions for the problems that were shown to be NP-complete. The solution space for a scheduling problem can be described by a tree of depth n, where depth k corresponds to the assignment of job k, for $1 \le k \le n$. Specifically, the root (depth 0) corresponds to an empty schedule - none of the jobs were assigned; at level 1 there are m nodes, representing job 1 being assigned to each of the m machines. At level k there are m^k nodes, corresponding to all possible assignment of the first k jobs. This implies that the brute-force solver may need to consider m^n assignments find the optimal one.

We note that, as detailed in Sect. 1.1, once the partition of jobs among the machines is determined, the internal job order on each machine either has no effect on the solution (in the min C_{max} problem), or is the unique SPT-order (in the min $\sum_j C_j$ problem). Thus, the solution space need not distinguish between assignments with different internal order of the same set of jobs on every machine.

Obviously, even without considering different internal orders, iterating over all of the m^n configurations is not feasible when dealing with large instances. Our solver uses a *branch and bound* technique combined with other optimizations to effectively trim tree-branches that are guaranteed not to yield an optimal solution.

In particular, the solver keeps in memory the best solution it found so far (its objective value and its transition cost). When processing a tree node if the already accumulated transition cost is larger than the budget or if the objective-function value is larger than the current best, then the solver can safely discard this tree branch as it is guaranteed not to yield a feasible optimal solution. For partial assignments, the objective-value is calculated by combining the value (makespan or total flow-time) of the already assigned jobs, and a lower bound on the yet-to-be-assigned jobs. For the minimum makespan problem, the lower bound is calculated by assuming perfect load-balancing ($\sum_j p_j / m$), and for total flow-time the lower bound is calculated by assuming SPT-order with no job-extension penalties.

In addition, we find out that considering the jobs from longest to shortest, that is, depth 1 corresponds to the longest job in the initial assignment, etc.) drastically helps in trimming branches earlier in the process.

The solver was designed to use multi-core machines in order to shorten the run time, by doing the work in parallel on the different cores. In the heart of the design stands a concurrent queue to which tasks are enqueued. Different threads concurrently dequeue these tasks, in a consumer-producer like mechanism. A 'task' for that matter is a request to process a tree node. That is, when the solver starts the queue is empty and a task to process the root node is added, which ignites the process. The solver is done when the queue is empty.

The solver's ability to solve problems instances of different sizes is determined by the given machine, to be more specific, by the CPU's clock speed, the number of available cores and sufficient memory (as the entire process is in memory). For example, the solver was able to handle a problem with 9 machines and 20 jobs in about 100 minutes, when ran on a machine with 8 cores and 32GB of RAM memory.

4.2 An Optimal Algorithm for $\Sigma_j C_j$

An algorithm for finding an optimal reschedule with respect to the minimum total flow-time objective is presented in [2]. The algorithm returns an optimal modified schedule using the minimal possible budget. The algorithm is based on reducing the problem to a minimum-weight complete-matching problem in a bipartite graph. The algorithm fits the most general case - arbitrary modifications, arbitrary transition costs and arbitrary job-extension penalties.

We have implemented this algorithm, and use its results as a benchmark so we can evaluate how well our heuristics perform. The algorithm is based on matching the jobs with possible slots on the machines. For completeness, we give here the technical details that are relevant to its implementation. Recall that n and m represent, respectively, the number of jobs and machines in the modified instance. Let $G = (V, E)$, where $V = J \cup U$. The vertices J correspond to the set of n jobs (a single node per job). The set U consists of mn nodes, q_{ik}, for $i = 1, \ldots, m$ and $k = 1, \ldots, n$, where node q_{ik} represents the kth from last position on machine i. The edge set E includes an edge (v_j, q_{ik}) for every node in J and every node in U (a complete bipartite graph). The edge weights consist of two components: a dominant component corresponding to the contribution of a job assigned in a specific position to the total flow-time, and a minor component corresponding to the associated transition cost. Both components are combined to form a single weight. Formally, for a large constant Z,

- For every job that is assigned to i in S_0, let $w(v_j, q_{ik}) = Zkp_j$.
- For every $i' \neq i$, let $w(v_j, q_{i'k}) = Zk(p_j + \delta_{i,i',j}) + \theta_{i,i',j}$.

These weights are based on the observation that a job assigned to the k-th from last position, contributes k times its processing-time to the total flow-time (see details in [2]). We implemented the algorithm by using the Hungarian method [19], a combinatorial optimization algorithm that solves the assignment problem in polynomial time. The solver's run time is $O(|V|^3)$, where $|V| = n(m + 1)$. In practice, this optimal solver can easily handle instances with 30 machines and 300 jobs.

5 Our Heuristics

In this section we describe the heuristics we have designed and implemented. Some heuristics were designed for specific modification (e.g. machines removal, limited budget), or for specific objective function, while some are general and fit all our reoptimization variants.

5.1 Heuristics for Minimum Makespan

We suggest two greedy heuristics, both intended to solve the *minimal makespan* problem (min C_{max}). In the first, we select the next migration to be performed according to the job's processing times, while in the second, we select the next migration according to the loads on the machines. Both algorithms begin with S_0 as the initial configuration. If the modification involves machines' removal, we first perform migrations of jobs assigned to the removed machines and migrate each such job j assigned to a removed machine i, to a machine $i' \in M$ for which $\theta_{i,i',j}$ is minimal. Ties are broken in favor of short extension penalty. As mentioned in the introduction, we assume that the budget is sufficient for this reschedule, as otherwise no feasible solution exists. Following the above preprocessing, we perform the following:

1. *LPT-Based:* In every iteration we consider the jobs in non-increasing processing-time order. When considering job j, we check whether migrating it to one of the two least loaded machines increases the load-balancing, formally, assume j is assigned to machine i, and we consider moving it to machine i', we check whether $p_j + \theta_{i,i',j} + L_{i'} < L_i$. If the answer is positive and the remaining budget allows, the migration is performed. We repeat the iterations until a complete pass over the jobs yields no migration.

2. *Loads-Based:* In every iteration we try to migrate some job out of the most loaded machine. We first consider the pair of most-loaded and least-loaded machines. Denote these machines by i and i'. We consider jobs on machine i according to order $\theta_{i,i',1} \leq \theta_{i,i',2} \leq \dots$. When considering job j we check whether migrating it to machine i' increases the load-balancing, that is, $p_j + \theta_{i,i',j} + L_{i'} < L_i$. If the answer is positive and the remaining budget allows, the migration is performed, and a new iteration begins (maybe with a different pair of most- and least-loaded machines). If the answer is negative for all the jobs on i, we move to the next candidate for target machine i' - the second least-loaded machine, etc. The iteration ends when some beneficial migration from the most-loaded machine is detected. If none such migration exists, the algorithm terminates.

5.2 Heuristics for Minimum Total Flow-Time

The minimum total flow-time reoptimization problem can be solved optimally assuming unlimited budged. While the optimal algorithm presented in Sect. 4.2 solves optimally the $\Sigma_j C_j$ problem using the minimal possible budget, it cannot be modified to solve the problem when the budget is limited. In fact, as shown in Sect. 2, this variant is NP-hard. We propose two heuristics that use the optimal algorithm as a first step and then each, in its own way, change the assignment to reach a feasible solution which obeys the budget constraints. A third algorithm that we propose, tries to reach an SPT-like schedule.

1. Greedy Reversion: The optimal algorithm returns an assignment S minimizing the total flow-time, which might not conform to the budget limitation. The following steps are performed to reach a feasible solution. First, we sort all the jobs which migrated in the transition from S_0 to S in non-increasing order according to the transition cost their migration caused.

We then distinguish between two cases:

1. The modification consists of only machines' *addition*. We revert the transitions one by one until we reach an allowed budget.
2. The modification includes machines' *removal*. We revert the transitions of jobs which do not originate from a removed machine, one by one until we reach an allowed budget. If after all possible reverts were done, the budget is still not met, we continue to the next step: 'Handling jobs of removed machines'.

Handling jobs of removed machines: This step is performed only when removed machines are involved, and all the jobs assigned to remaining machines are back in their initial machines. We sort the jobs originated from removed machines in non-increasing order according to the transition cost their migration (determined by the optimal algorithm) caused. Job after job, we migrate a job j assigned in S_0 to a removed machine $i \in M_0 \setminus M$ to the machine $i' \in M$ for which $\theta_{i,i',j}$ is minimal, breaking ties in favor of better objective value. As explained in the introduction, we assume that the budget is sufficient to complete all these migrations.

2. Cyclic Reversion: The optimal algorithm returns an assignment S minimizing the total flow-time, which might not conform to the budget limitation. Similar to the previous heuristic, we choose the most expensive transition involved, denote by j the corresponding job. We migrate job j back to its origin machine, $M_{0,j}$. Since we wish to keep jobs distributed as evenly as possible, we now choose a job that migrated to $M_{0,j}$ and migrate it back to its initial machine. We choose the job whose migration to $M_{0,j}$ was most expensive. We keep these cyclic reverts until one of following conditions holds: (a) We made a complete loop and reached back the machine from which we started. (b) We have reached a machine to which no job was migrated. If the budget conforms to the limitation, we stop, Otherwise we choose a job with the most expensive transition cost and start a new revert cycle.

If the modification includes machines' *removal*, jobs originated from the removed machines cannot be selected to migrate back to their original machine. If the budget is not met after all the allowed reverts were performed, we continue to the step 'Handling jobs of removed machines', as described in the greedy heuristic.

3. SPT-like: It is known that SPT ordering is optimal for $P||\Sigma_j C_j$. We therefore try to reach a schedule that fulfills the following basic properties of an SPT schedule:

1. In any optimal schedule the number of jobs on any machine is either $\lfloor \frac{n}{m} \rfloor$ or $\lceil \frac{n}{m} \rceil$.
2. The jobs can be partitioned into $\lfloor \frac{n}{m} \rfloor$ rounds, such that the k-th round consists of all the jobs that are k from last on some machine. In an SPT schedule, each of the jobs in the k-th round is not shorter than each of the jobs in the $k + 1$st round.

This heuristic consists of three stages. The first stage, applied when machines were removed, is to move to some feasible schedule - each of the jobs assigned to a removed machine, is migrated into a machine for which the transition cost is the lowest.

In the second stage, we try to make the machines as balanced as possible in terms of number of jobs. While the budget allows and while there exists a pair of machines, one with more than $\lceil \frac{n}{m'} \rceil$ jobs and the other with less than $\lfloor \frac{n}{m'} \rfloor$ jobs, we migrate the cheapest-to-move job on the first machine, to the second one.

In the third phase, we try to make our solution as close to the SPT ordering as possible, we compare the solution round after round to the desired SPT ordering, and switch jobs whenever required, as long as the budget allows.

5.3 Genetic Algorithm

In the Genetic algorithm the idea is for the solution to be obtained in a way that resembles a biological evolution process [10, 22]. That is, we let the method's evolutionary process find the solution by itself. The idea is to define the *Genome* of a single solution and a *Ranking* method *Rank* : *Genome* $\rightarrow \mathbb{R}$. The genome of a single solution represents and gives all the needed information regarding the solution. In our case the Genome is simple, for a problem instance with m machines and n jobs, a solution genome id, g, defined as $g = (g_1, g_2, \ldots, g_n)$, where $g_i \in [1, m]$ and $g_i \notin \{x | x$ is a removed machine id$\}$, in other words each cell with index i represents the $i - th$ jobs and the cell's value is the machine this job is assigned to in the modified schedule. The ranking method is used to define how good is a given solution. In our case the ranking method sorts the different genomes first by the objective method value (C_{max} or $\Sigma_j C_j$) and then by the transition cost. We create a *population* (generation 1), which is a collection of *genomes*, we rank each member of the population and sort them from best to worst.

When solving a reoptimization problem with limited budget, to guarantee the algorithm end up with a feasible solution, we create at least one feasible genome in generation1. In the case of 'machines addition', this solution will be S_0 as it is both valid and requires no transition cost. In the case of 'machines removal', a job j assigned in S_0 to a removed machine $i \in M_0 \setminus M$ is assigned in the feasible genome to the machine $i' \in M$ for which $\theta_{i,i'j}$ is minimal. We assume that the budget is sufficient for this reschedule, as otherwise no feasible solution exists.

The next step is the evolutionary-like step, in which we create the next generation according to the following methods:

1. Elitism mechanism: We take the best 5% genomes and move them 'as-is' to the next generation. This guarantee that the next generation will be at least as good as the current one. To deal with the case of limited budget, we also pass 'as-is' the best solution which meets the given budget. This must be done since the genetic process is pushing the genomes population for a better objective values as a primary goal and to minimize the transition cost as a secondary goal.

2. Cross over: From the entire genomes population, we choose randomly 2 elements, denoted g_x and g_y, we choose a pivot point from the range of $ind \in [1, n-1]$, and create two new items:

$$newItem1_i = \begin{cases} g_{x_i} & \text{if } i \leq ind \\ g_{y_i} & \text{if } i > ind \end{cases}$$

and

$$newItem2_i = \begin{cases} g_{y_i} & \text{if } i \leq ind \\ g_{x_i} & \text{if } i > ind \end{cases}$$

43% of the next generation is a result of this operator.
3. Mutate: We choose a random genome, we choose from its genome a random cell and change its value. 43% of the next generation is a result of this operator.
4. Fresh Items: We generate new genomes. These new elements have the potential of shifting the results in a new direction and to help avoid local optimum. 9% of the next generation is be a result of this operator.

Each genome in the newly created population is then re-evaluated, meaning, its score is computed. The process repeats itself until it fails to improve any further and the genome with the best ranking is selected as output from the most recent generation. Obviously if we examine the best solution from generation $i + 1$ compared to that of generation i we will notice that the 'quality' of the best solution is non decreasing over the generations as we have the 'Elitism mechanism' which ensures us that the best individuals will survive to the next generation.

6 Experimental Results

The datasets for our experiments were created using our own data generator which supports any parameters combination of number of machines and jobs, number of added or removed machines, number of added or removed jobs, as well as the distributions of job lengths, job-extension penalty, and transition costs.

Instances on which we run our heuristics could be very large (hundreds of jobs, and several dozens of machines). Instances on which we run our brute-force solver (introduced in Sect. 4.1) had to be smaller. In particular, we ran the brute-force solver on instances with 20 jobs and 4 machines. We find out that even such small instances can provide a good comparison between different heuristics; therefore, the brute-force solver is helpful for concluding how far from the optimum our heuristics perform. The optimal algorithm for min $\Sigma_j C_j$, based on a reduction to perfect matching (introduced in Sect. 4.2), was able to handle relatively large instances of 15 machines and 300 jobs. In our basic template for job creation, the jobs' lengths were uniformly distributed in [1, 20]. The job-extension penalty of job j was uniformly distributed in [1, $\frac{p_j}{2}$], and the transition costs were uniformly distributed in [1, 5].

To generate problems instances for a specific experiment we took a template instance, decided on one parameter that will vary in the different runs, and set the rest of the parameters to basic values. For example, to understand how the number of added machines affects the heuristics performance, we fixed all the other parameters (jobs' lengths, transition costs, etc.), and run the heuristics on multiple instances which vary only in the number of added machines. To avoid anomalies, we have generated for each experiment 5 instances with the same parameters, based on 5 templates instances (same configuration, different instance) and considered the average performance as the result.

Another parameter that could affect the performance of our heuristics is the initial assignment - which may be random, SPT or LPT. We found out that in practice the initial assignment does not affect the results significantly and the results we present in the sequel were all generated with a randomize initial assignment - which is a bit more 'challenging' and therefore emphasizes the differences among the heuristics.

The results of our experiments are presented and analyzed below. In all the figures, the bars show the objective value ($\Sigma_j C_j$ or C_{max}), and the lines show the corresponding transition cost.

6.1 Results for the Minimal Total Flow-Time Problem

6.1.1 Machines' Addition

The template for heuristics that analyze the min $\Sigma_j C_j$ problem consists of 15 machines and 300 jobs. We start by showing how the different heuristics performs on instances with both transition costs and job-extension penalties, where the number of added machines was set to $m/2$. As shown in Fig. 2, with unlimited budget the genetic algorithm is the closest to the known optimum, calculated by the perfect matching algorithm result. As budget is limited, the performance of the genetic algorithm drops. As expected, the lower the budget, the higher the objective value. Also, the differences between the heuristics are less significant as the budget decreases, as less transitions are allowed. Interesting fact is that the genetic algorithm has a slight improvement as the limitation is getting stricter. We explain this by the fact that before starting the algorithm we include in the population items that obey the allowed budget. Later, these items are influencing the genetic process and are helping the algorithm to converge to a good solution, better than with a more relaxed budget limitation.

The goal of our next experiment was to see how close the heuristics get to the actual optimum. For this test we have used our brute-force solver on relatively small problem instances (4 machines and 20 jobs), two machines were added and the budget was set to 20. The results for various extension penalties are shown in Fig. 3. We observe that both 'Greedy-reversion' and 'Cyclic-reversion' heuristics were very close to the optimum.

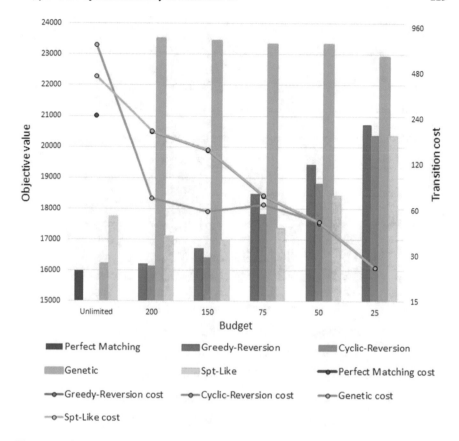

Fig. 2 Results for min $\Sigma_j C_j$ with $m/2$ added machines and variable budget

6.1.2 Machines' Removal

Figure 4 presents the performance of the different heuristics when the modification is machines' removal and the budget is limited to 150. According to our parameters, this budget is expected to be sufficient for the migration of about 20% of the jobs. The initial assignment was of 300 jobs on 15 machines. Not surprisingly, we see an increase of the objective value as the number of removed machines increases. All of the heuristics performed more or less the same, both in terms of the achieved objective value and in term of the budget utilization, with an exception of the Genetic algorithm which manage to use a significantly lower budget.

Figure 5 presents the results for the same instance and the same modification only with unlimited budget. This problem is the one for which we have an efficient optimal algorithm (see Sect. 4.2). The genetic algorithm perform very close to the optimum for every number of removed machines. In fact, for two removed machines its transition cost is lower than the optimum and only slightly higher in the total

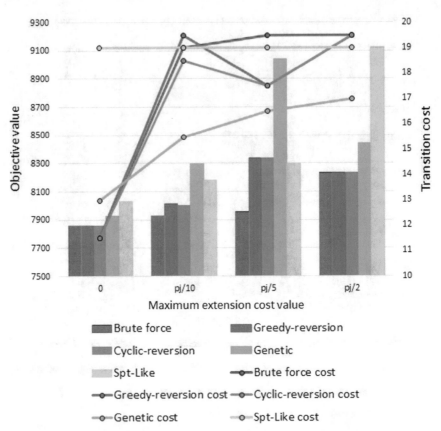

Fig. 3 Results for min $\Sigma_j C_j$ with variable extension penalty

flow-time (recall that the optimal algorithm 'insists' on finding a reschedule that minimizes the total flow-time). On the other hand the SPT-Like heuristic is both very costly and gives poor results. This can be explained by the fact that insisting on a complete SPT order requires many transition, and involves many job-extension penalties.

6.2 Results for the Minimum Makespan Problem

6.2.1 Machines' Addition

Our template for experiments analyzing the min C_{max} problem consists of 30 machines and 500 jobs. Figure 6 presents results for adding 15 machines and variable budget. The 'Loads-based' heuristic is the best heuristic. The genetic algorithm

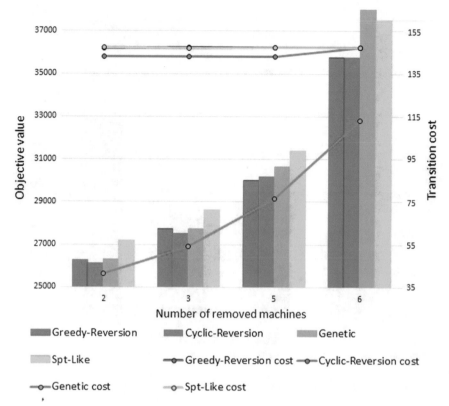

Fig. 4 Results for $\Sigma_j C_j$ for machines' removal and limited budget

perform poorly compared to the other heuristics, but on the other hand, it does not utilize the whole budget.

In the our next experiment we measure how close the heuristics get to the optimum. We have used our brute-force solver on relatively small problem instances, consisting of 4 machines, 20 jobs, and a limiting budget of 20. The results for various extension penalties are shown in Fig. 7. Once again, the 'Loads-based' heuristic outperform the others, and is relatively close to the optimum. The 'LPT-based' heuristic seems to perform (relatively) better as the job-extension penalty increases.

6.2.2 Removing Machines

Our next experiments compare the performance of the different heuristics when the modification is machines' removal. We performed two experiments - with budget limited to 250 and with unlimited budget. The results are shown in Figs. 8 and 9

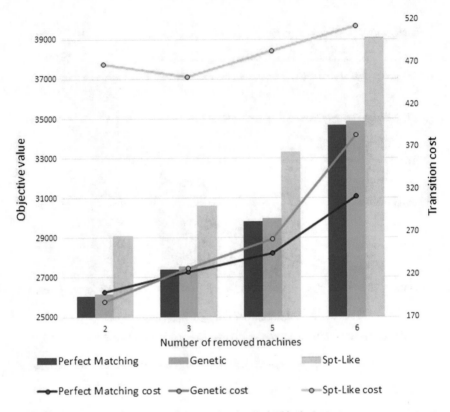

Fig. 5 Results for $\Sigma_j C_j$ for machines' removal and unlimited budget

respectively. Our results show that with unlimited budget, all three heuristics (LPT-based, Loads-based and genetic) perform more or less the same. While a similar makespan is achieved, the Loads-based heuristic requires the lower transition cost, then the LPT-based (that needs 10% higher cost) and the genetic ($15 - 20\%$ higher than Loads-based). With limited budget, the Loads-based and LPT-based heuristics perform significantly better than the genetic algorithm, but they also require a much higher budget.

7 Conclusions and Future Work

We presented theoretical and experimental results for the reoptimization variant of job scheduling problems. We have shown that the problems of finding the minimum total flow-time with limited budget, finding the minimum makespan with limited budget

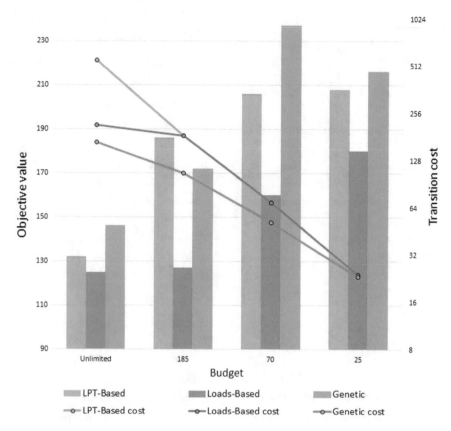

Fig. 6 Results for min C_{max} with $m/2$ added machines and variable budget

and finding the minimum makespan with unlimited budget are NP-Complete. We have designed and implemented several heuristics for each of these problems, and performed a comprehensive empirical study to analyze their performance. To see how well these heuristics perform compared to the actual optimum, an efficient branch-and-bound brute-force solver was designed and implemented. An optimal algorithm for the minimum total-flow problem with unlimited budget was also implemented.

In general, our experiments reveal that while the problems are NP-hard, heuristics whose time complexity is polynomial in the number of jobs and machines, perform very well in practice. Simple algorithms, that are based on adjustment of known heuristics for the one-shot problem (with no modifications) are both simple to implement and provide results that are, on average, within $10 - 15\%$ from the optimum. More complex algorithms, that are based on a preprocessing in which a perfect matching algorithm is implemented, perform on average even better.

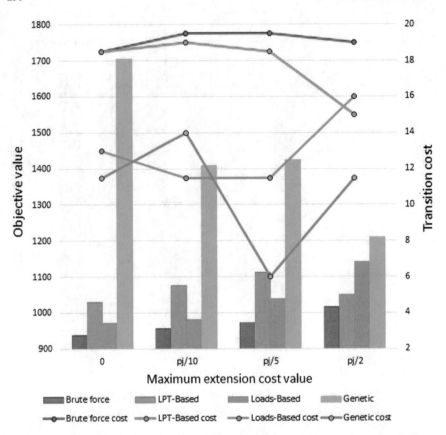

Fig. 7 Results for min C_{max} with variable extension penalty

We have observed that while the Genetic algorithm does not perform well when given a limited budget, it performs relatively well with unlimited budget for the minimum C_{max} problem, and close to optimum for $\Sigma_j C_j$. It also takes a considerable amount of time to run. In some scenarios, its objective value may not be competitive compared to other heuristics, however, its budget utilization is impressively good (see for example Fig. 4). A known issue of genetic algorithm is that the parameters must be carefully tuned in order for the algorithm to converge into a good solution. Future work on this algorithm might refactor the population size or operators we have used (Elitism, Crossover, Mutation) by adding new operators, modify the existing or change the possibilities of each to create a more optimized genetic algorithm. Another direction to explore is to create a dedicated score method for each variant of the problems. For real life applications where budget utilization is a big concern, we are sometimes allowed to use lot of budget but strive to use as less as possible. The genetic algorithm is a good choice for such scenarios.

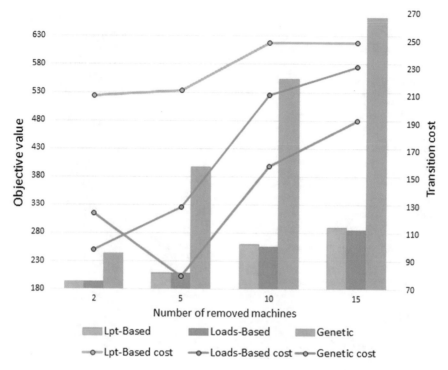

Fig. 8 Performance for min C_{max} for machines' removal and limited budget

Our greedy heuristics performed well, both on the $\Sigma_j C_j$ and the C_{max} problems. We have observed that the 'Loads-based' heuristic outperformed the 'LPT-based' both in terms of objective value and budget utilization. With unlimited budget, the budget utilization difference was more significant. For min $\Sigma_j C_j$, the 'Cyclic-reversion' showed better performance compared to the 'Greedy-reversion', This result does not surprise us as a more balanced solution is expected to yield better results for the minimum total flow-time problem. In terms of budget utilization, it was expected that this greedy, budget oriented method will utilize as much budget as possible.

An additional direction for future work is to develop algorithms for the minimum makespan problem with a guaranteed approximation ratio. While tuning existing approximation-algorithm for the classical one-shot problem seems to be a promising direction, the presence of transition-costs and job-extension penalties give rise to new challenges and considerations. Finally, it would be interesting to consider different objective functions, or different scheduling environments, for example, jobs with deadlines or precedence constraints, as well as unrelated or restricted machines.

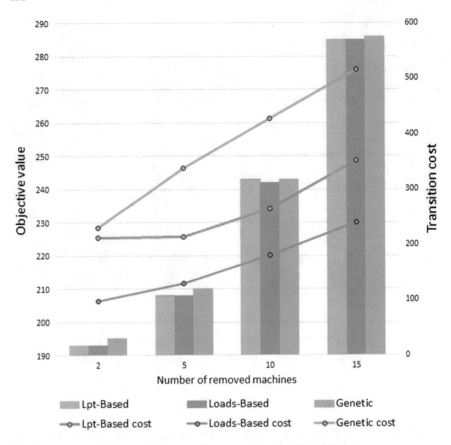

Fig. 9 Performance for min C_{max} for machines' removal and unlimited budget

References

1. Ausiello, G., Escoffier, B., Monnot, J., Paschos, VTh: Reoptimization of minimum and maximum traveling salesmans tours. J. Discret. Algorithms **7**(4), 453–463 (2009)
2. Baram, G., Tamir, T.: Reoptimization of the minimum total flow-time scheduling problem. Sustain. Comput. Inf. Syst. **4**(4), 241–251 (2014)
3. Belikovetsky, S., Tamir, T.: Load rebalancing games in dynamic systems with migration costs. Theor. Comput. Sci. **622**(4), 16–33 (2016)
4. Berlinskaa, J., Drozdowskib, M.: Scheduling divisible MapReduce computations. J. Parallel Distrib. Comput. **71**(3), 450–459 (2011)
5. Birrell, A.D., Nelson, B.J.: Implementing remote procedure calls. ACM Trans. Comput. Syst. **2**, 39–59 (1984)
6. Bockenhauer, H.J., Forlizzi, L., Hromkovic, J., Kneis, J., Kupke, J., Proietti, G., Widmayer, P.: On the approximability of TSP on local modifications of optimally solved instances. Algorithmic Oper. Res. **2**(2) (2007)
7. Bruno, J.L., Coffman, E.G., Sethi, R.: Scheduling independent tasks to reduce mean finishing time. Commun. ACM **17**, 382–387 (1974)

8. Clark, C., Fraser, K., Hand, S., Hansen, J.G., Jul, E., Limpach, C., Pratt, I., Warfield, A.: Live migration of virtual machines. In: The 2nd Symposium on Networked Systems Design and Implementation (NSDI) (2005)
9. Conway, R.W., Maxwell, W.L., Miller, L.W.: Theory of Scheduling. Addison Wesley, Reading (1967)
10. Eiben, A.E., Smith, J.E. Introduction to Evolutionary Computing. Springer, Berlin (2007)
11. Epstein L., Levin, A..: Scheduling with processing set restrictions: PTAS results for several variants. Int. J. Prod. Econ. **133**(2) (2011)
12. Eppstein, D., Galil, Z., Italiano, G.F.: Dynamic graph algorithms, Chap. 8. In: Atallah, M.J. (ed.) CRC Handbook of Algorithms and Theory of Computation (1999)
13. Escoffier, B., Milanic, M., Paschos, V.Th.: Simple and fast reoptimizations for the Steiner tree problem. DIMACS Technical Report 2007-01
14. Garey, M.R., Johnson, D.S.: Computers and Intractability: A Guide to the Theory of NP-completeness. W. H. Freeman and Co., New York (1979)
15. Graham, R.L., Lawler, E.L., Lenstra, J.K., Rinnooy Kan, A.H.G.: Optimization and approximation in deterministic sequencing and scheduling: a survey. Ann. Discret. Math. **5**, 287–326 (1979)
16. Hacking, S., Hudzia, B.: Improving the live migration process of large enterprise applications. In: The 3rd International Workshop on Virtualization Technologies in Distributed Computing (VTDC) (2009)
17. Hochbaum, D.S., Shmoys, D.B.: Using dual approximation algorithms for scheduling problems: practical and theoretical results. J. ACM **34**(1), 144–162 (1987)
18. Horn, W.: Minimizing average flow-time with parallel machines. Oper. Res. **21**, 846–847 (1973)
19. Kuhn, H.W.: The Hungarian Method for the assignment problem. Naval Res. Logist. Q. **2**, 83–97 (1955)
20. Leung, J.Y.-T., Li, C.-L.: Scheduling with processing set restrictions: a survey. Int. J. Prod. Econ. **116**(2), 251262 (2008)
21. Parallels Virtuozzo. http://www.parallels.com/products/pvc
22. Schmitt, L.M.: Theory of genetic algorithms. Theor. Comput. Sci. **259**, 1–61 (2001)
23. Shachnai, H., Tamir, G., Tamir, T.: Minimal cost reconfiguration of data placement in storage area network. Theor. Comput. Sci. **460**, 42–53 (2012)
24. Schieber, B., Shachnai, H., Tamir, G., Tamir, T.: A Theory and Algorithms for Combinatorial Reoptimization. Algorithmica, to appear
25. Smith, W.E.: Various optimizers for single-stage production. Naval Res. Logist. Q. **3**, 59–66 (1956)
26. Thorup, M., Karger, D.R.: Dynamic graph algorithms with applications. In: Proceedings of 7th SWAT (2000)
27. Xen Project. http://www.xenproject.org/

Author Index

© Springer International Publishing AG 2018
S. Fidanova (ed.), *Recent Advances in Computational Optimization*,
Studies in Computational Intelligence 717, DOI 10.1007/978-3-319-59861-1

Printed in the United States
By Bookmasters